TSUNAMI HAZARD

A Practical Guide for Tsunami Hazard Reduction

Edited by

E. N. BERNARD
Pacific Marine Environmental Laboratory, NOAA, Seattle, Washington, U.S.A.

Reprinted from *Natural Hazards*, Vol. 4, Nos. 2 & 3 (1991)

Kluwer Academic Publishers
Dordrecht / Boston / London

Library of Congress Cataloging-in-Publication Data

Tsunami hazard : a practical guide for tsunami hazard reduction /
edited by E.N. Bernard.
p. cm.
Selected papers from the 14th International Tsunami Symposium,
held July 31-Aug. 3, 1989, in Novosibirsk, U.S.S.R.; with additional
reports.
ISBN 0-7923-1174-4 (HB : acid free paper)
1. Tsunamis--Congresses. I. Bernard, E. N. (Eddie N.)
II. International Tsunami Symposium (14th : 1989 : Novosibirsk,
R.S.F.S.R.)
GC219.T76 1991
551.47'024--dc20 91-2137

ISBN 0-7923-1174-4

Published by Kluwer Academic Publishers,
P.O. Box 17, 3300 AA Dordrecht, The Netherlands.

Kluwer Academic Publishers incorporates
the publishing programmes of
D. Reidel, Martinus Nijhoff, Dr W. Junk and MTP Press.

Sold and distributed in the U.S.A. and Canada
by Kluwer Academic Publishers,
101 Philip Drive, Norwell, MA 02061, U.S.A.

In all other countries, sold and distributed
by Kluwer Academic Publishers Group,
P.O. Box 322, 3300 AH Dordrecht, The Netherlands.

Printed on acid-free paper

Printed in the Netherlands

Table of Contents

Preface

The Fourteenth International Tsunami Symposium held from 31 July to 3 August 1989 in Novosibirsk, U.S.S.R., was sponsored by the International Union of Geodesy and Geophysics. Sixty-five scientists from 13 countries met to exchange information on recent advances in tsunami research. The Symposium was a great success due to the enthusiasm of the participants, the quality of research presented, and the great organization provided by the Soviet hosts. Teams of dedicated workers, under the fine leadership of Academician A. S. Alexseev and Dr V. K. Gusiakov, blended social and scientific activities in a memorable fashion.

The 62 presentations of the Symposium were divided into six areas of research: generation (7), propagation (12), coastal effects (10), observations (11), seismics and tectonics (10), and hazard mitigation (12). A summary of the research presented appears as the first article in this special issue. Following the Symposium, a team of session chairmen nominated 20 of these oral presentations to be published in a special issue devoted to the International Tsunami Symposium.

Of the 20 nominations, 13 manuscripts from seven countries were accepted for publication in this issue. The scientific reports are grouped into three areas of research: observations (beginning with Gonzalez *et al.*), physical processes (beginning with Shuto), and hazard mitigation (beginning with Tinti). Included in this special issue also are meeting reports from the Tsunami Workshop and the Intergovernmental Oceanographic Commission's Twelfth Session of the International Coordinating Group for the Tsunami Warning System in the Pacific. Also included is the Opening Address of the Symposium in which tsunami research is reviewed over the past 30 years and new opportunities are identified.

In short, this issue represents highlights of Tsunami 89. I acknowledge the support and generosity of the authors and 39 reviewers, who set high scientific standards, and to Lenora Cahoon, whose competent assistance made this project a success. Finally, I thank Professor El-Sabh whose encouragement and persistence made this issue possible.

E. N. BERNARD
Guest Editor

Natural Hazards **4**: 115–117, 1991.

Fourteenth International Tsunami Symposium: Opening Address

On behalf of the Tsunami Commission, I have the great privilege of opening the International Tsunami Symposium of the International Union of Geodesy and Geophysics here in Novosibirsk, U.S.S.R. It is my great honor to extend to all of you a very cordial welcome to this symposium. First of all, as the Chairman of the Tsunami Commission, I wish to express my heartfelt gratitude to the Soviet National Committee for the organization of International Tsunami Symposium 1989 and, in particular, to Professor A. S. Alexeev, who led the committee in its untiring efforts in hosting this symposium.

The Tsunami Commission of the International Union of Geodesy and Geophysics (IUGG) was created in 1960 in Helsinki, Finland, at the 12th General Assembly to promote the exchange of scientific and technical information about tsunamis among nations concerned with the tsunami hazard. Since its beginning, the Commission has sponsored 14 tsunami symposia and has encouraged the publication of 10 proceedings containing over 300 research reports from these symposia. Since 1960, tsunami symposia have been held throughout the world as can be seen by this list of dates, places, and proceeding publications.

Tsunami Symposia Sponsored by the Tsunami Commission

1961 Honolulu, Hawaii – D. C. Cox (U.S.A.), editor
1963 Berkeley, California
1966 Berne, Switzerland
1969 Honolulu, Hawaii – W. M. Adams (U.S.A.), editor
1971 Moscow, U.S.S.R. – S. L. Soloviev (U.S.S.R.), editor
1974 Wellington, New Zealand – R. A. Heath and M. M. Cresswell (New Zealand), editors
1975 Grenoble, France
1977 Ensenada, Mexico – T. S. Murty (Canada), editor
1979 Canberra, Australia – R. D. Braddock (Australia), editor
1981 Sendai, Japan – K. Iida and T. Iwasaki (Japan), editors
1983 Hamburg, Germany – E. N. Bernard (U.S.A.), editor
1985 Victoria, Canada – T. S. Murty and W. J. Rapatz (Canada), editors
1987 Vancouver, Canada – E. N. Bernard (U.S.A.), editor
1989 Novosibirsk, U.S.S.R

Tsunami scientists are proud of these past accomplishments, but new challenges are now emerging on the horizon. As the Tsunami Commission stands on the threshold of its fourth decade, the 1990s offer a unique challenge – the International Decade of Natural Disaster Reduction. For 30 years, tsunami scientists have worked toward the goal of reducing the hazard of tsunami. To place those years in perspective, let us briefly examine the history of tsunami research (1960–1989) by decade. After examining the past, I would like to discuss the challenge of the future.

The 1960s – Discovery

The sixties were a decade of defining the tsunami phenomenon. The tsunamis of 1960 and 1964 convinced everyone that Pacific coastal residents must be protected from this hazard. Ray-tracing techniques along with analytical studies yielded important information about travel times and physical characteristics of tsunami. Physical models were used to evaluate the influence of protective barriers and to determine inundation zones. Several instruments were tested for use in detecting tsunamis including ocean acoustic waves, atmospheric waves, and bottom pressure gages. This was an exciting time in tsunami research, as many discoveries were made in this fertile field.

The 1970s – Computer Applications

In the seventies we saw the emergence of numerical models which were based on the analytical and physical models of the sixties. Extensive numerical modeling experiments were conducted on tsunami generation, propagation, and run-up for local, regional, and Pacific-wide areas. No major Pacific-wide tsunami occurred during this decade, so interest in the problem started to fade.

The 1980s – Hazard Applications

In the eighties tsunami scientists began the process of converging these theoretical principles for use in hazard management. The decade of the 1970s provided a set of useful numerical models to conduct scenario experiments. These numerical models, however, could not be properly verified because of a lack of data. Not only were tsunami data few, but the type available were not suited for model validation. During the 1981 Sendai Tsunami Symposium, Professor Iida stated in his opening address:

> Numerical calculations of tsunami problems have advanced considerably. It is said, however, that systematic measurement of tsunamis in the open ocean for research purposes and for reliable tsunami warning is still not accomplished. Further technical improvements are necessary for fast and efficient tsunami warning, and tsunami risk estimations are required for the prevention of tsunami hazards to coastal inhabitants and important construction in coastal regions of the Pacific and other oceans. To cover these requirements, we need further various tsunami data and their analysis.

These words were rather prophetic in that the 1983 Sea of Japan tsunami killed over 100 Japanese citizens. With this natural reminder of the tsunami hazard, scientists have made great progress in all three areas. The technology now exists for measuring tsunamis in the deep ocean, historical data have been carefully collected and analyzed for hazard assessment in several countries, and a fast and efficient local warning system using satellites has been developed. One important result of the historical data studies is the emerging need for tsunami hazard reduction outside the Pacific Ocean. During this symposium, these three topics, as well as other important research, will be discussed in detail.

The 1990s – Hazard Reduction

Now let us turn to the future. As I said at the beginning of this discussion, the 1990s pose a special challenge because of the international effort in disaster reduction. I believe tsunami scientists are willing and able to rise to the challenge of disaster reduction. I further believe that the tsunami community has the necessary tools to focus on a single objective that would do much to reduce the tsunami hazard for our global community. That single objective would be an '*internationally accepted methodology for preparing tsunami flooding maps*'.

I would hope that by 1999, we could characterize the 1990s as the Decade of Disaster Reduction. I would further hope that scientists would be reporting on the recent advances of tsunami inundation mapping that yield accuracies of 20%. In order to accomplish this goal, scientists must adopt a process in which several tsunamis are carefully measured. Such a process would include observations of the generation and propagation (in oceanic depths from 50 to 5000 m) of a tsunami and measurements of the flooding along the coastline. International cooperation will be required to instrument probable source and affected areas.

A second component of the process will be a focused modeling effort. Again, models must reasonably simulate tsunami generation and propagation in order to accurately define areas that will be flooded. International coordination is required to meet the needs of affected nations. Models should be developed on personal computers to facilitate technology transfer.

The tsunami scientific community has all these techniques/tools in place. A combined effort is required to marshal the necessary political, fiscal, and intellectual resources of all nations to make this a reality. The cost will be high. It will require activism on the part of tsunami scientists. It will require dedication on the part of emergency management. It will require resources that are not available today. As I said – it *is* a challenge – but one for which we have prepared for 30 years.

I sincerely hope that the Novosibirsk Tsunami Symposium will be remembered as the acceptance of this challenge.

Спасибо!

E. N. BERNARD
July 31, 1989

Natural Hazards **4**: 119–139, 1991.

The 1987–88 Alaskan Bight Tsunamis: Deep Ocean Data and Model Comparisons

F. I. GONZALEZ[1], C. L. MADER[2], M. C. EBLE[1], and E. N. BERNARD[1]
[1]*Pacific Marine Environmental Laboratory, NOAA, 7600 Sand Point Way, N.E., Seattle, WA 98115, U.S.A.*
[2]*Joint Institute for Marine and Atmospheric Research, 1000 Pope Road, Honolulu, HI 96822, U.S.A.*

(Received: 22 April 1990; revised: 30 July 1990)

Abstract. Excellent deep ocean records have been obtained of two tsunamis recently generated in the Alaskan Bight on 30 November 1987 and 6 March 1988, providing the best available data set to date for comparison with tsunami generation/propagation models. Simulations have been performed with SWAN, a nonlinear shallow water numerical model, using source terms estimated by a seafloor deformation model based on the rectangular fault plane formalism. The tsunami waveform obtained from the model is quite sensitive to the specific source assumed. Significant differences were found between the computations and observations of the 30 November 1987 tsunami, suggesting inadequate knowledge of the source characteristics. Fair agreement was found between the data and the model for the first few waves of the 6 March 1988 tsunami. Model estimates of the seismic moment and total slip along the fault plane are also in fair agreement with those derived from the published Harvard centroid solution for the 6 March 1988 event, implying that the computed seafloor deformation does bear some similarity to the actual source.

Key words. Tsunamis, ocean measurements, model comparisons.

1. Introduction

For decades, oceanographers have attempted to acquire deep ocean tsunami measurements for comparison with theoretical and numerical models. The need for high quality deep ocean data, free from the complicating effects of coastal resonance phenomena, has long been recognized (Bernard and Goulet, 1981; Raichlen, 1985). But making such measurements presents significant technical challenges, and earlier efforts have provided only two deep ocean bottom pressure recorder (BPR) records which clearly and unambiguously delineate an open ocean tsunami waveform. The first was acquired on 17 March 1979, 1000 km from the epicenter of a 7.6 M_s earthquake off the southwest Mexican coast (Filloux, 1982); the second was obtained on 19 June 1982, 1800 km from the epicenter of an M_s 6.9 earthquake off the Guatemala–El Salvadoran coast (Bernard and Milburn, 1985).

Advances in instrumentation have now made practical the establishment of deep ocean tsunami monitoring networks based on long-term (> 1 year) deployments of BPRs. Consequently, the Pacific Tsunami Observational Program (PacTOP) was recently initiated by the U.S. Department of Commerce's National Oceanic and

Atmospheric Administration (NOAA) to establish such a network; this effort reached a major milestone in the summer of 1986 with the establishment of five deep ocean BPR stations in the northeast Pacific. The stations are strategically located with respect to the Shumagin seismic gap, a region near the Aleutian Trench known to have seismic and tsunamigenic potential (Gonzalez *et al.*, 1987a, b).

In the second year of PacTOP, three sizable earthquakes occurred in the northeast region of the Alaskan Bight on 17 and 30 November 1987 and on 6 March 1988 (Lahr *et al.*, 1988). Each earthquake generated a small tsunami which was detected by tide gauges along the coastlines of Alaska, British Columbia, and the Hawaiian Islands. More importantly, each tsunami was also recorded by one or more BPRs in the PacTOP network. Within that 4-month period, the total number of deep ocean tsunami measurements acquired in the last decade was thereby increased from two to nine.

This report focuses on a comparison of long-wave numerical model simulations with the deep ocean BPR measurements of the two tsunamis which occurred on 30 November 1987 and 6 March 1988. The 17 November 1987 earthquake was the least energetic of the three and the resultant tsunami, which was unambiguously detected only at Yakutat and WC9, was relatively inconsequential and will not be dealt with here.

2. Tsunami Observations

Tsunami observations at the BPRs and several northeast Pacific tide gauge stations are presented in Figures 1 and 2 for 6 March 1988 and 30 November 1987, respectively. In this study we are primarily concerned with the BPR records, and certain features of these data which are particularly important to our comparison with the numerical simulations. The tide gauge records are presented here for the sake of completeness, but we will not attempt to compare numerical results with these data because they are strongly influenced by local bathymetry on a spatial scale which the model does not resolve. Specific examples of these local resonance effects in the tide gauge tsunami records and a general discussion of all of the deep ocean and coastal tsunami observations, including those acquired in Hawaii, will be provided in a later report.

Briefly, the tsunami waveforms recorded at abyssal stations AK7 and AK8 (Figures 1 and 2) are classically dispersive and amplitude-modulated to form distinct wave packets, as predicted by theory (see, e.g., Kajiura, 1963, or Mei, 1983). In spite of relatively high background noise levels, these same features are also apparent in the 6 March tsunami record at station AK10 (Figure 1) and in both tsunami records acquired at station WC9 (Figures 1 and 2).

The higher background energy levels which characterize the records at WC9 and AK10 are probably due to the station locations. WC9 is situated in the Axial Caldera on the mid-oceanic Juan de Fuca ridge, and AK10 is located on the landward slope of the Aleutian Trench. Both are therefore in much shallower water

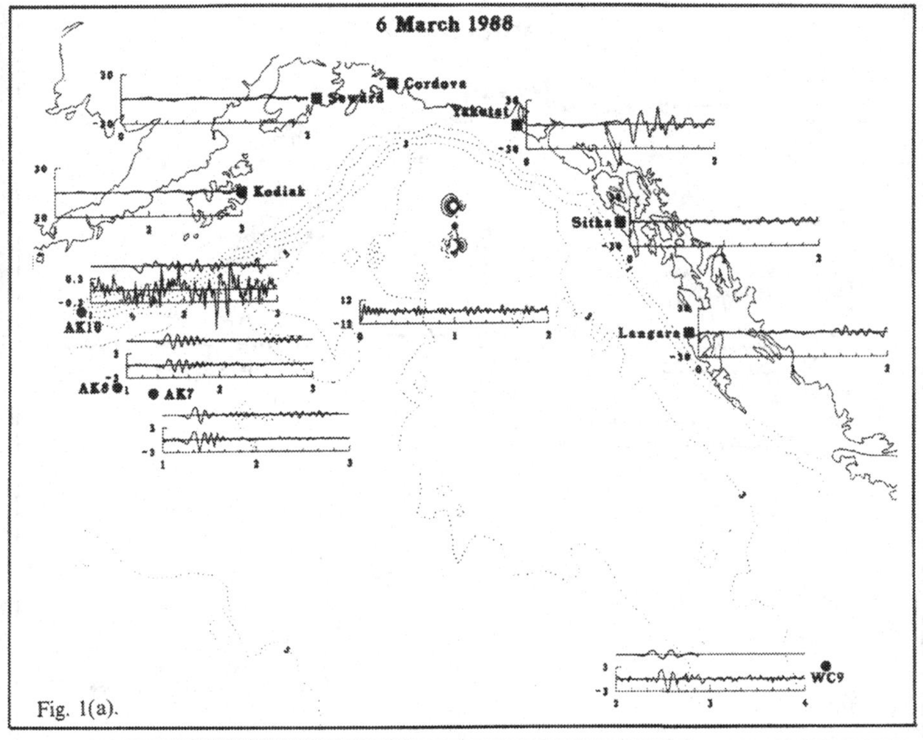
6 March 1988
Cordova
Seward
Yakutat
Kodiak
Sitka
Langara
AK10
AK8
AK7
WC9
Fig. 1(a).

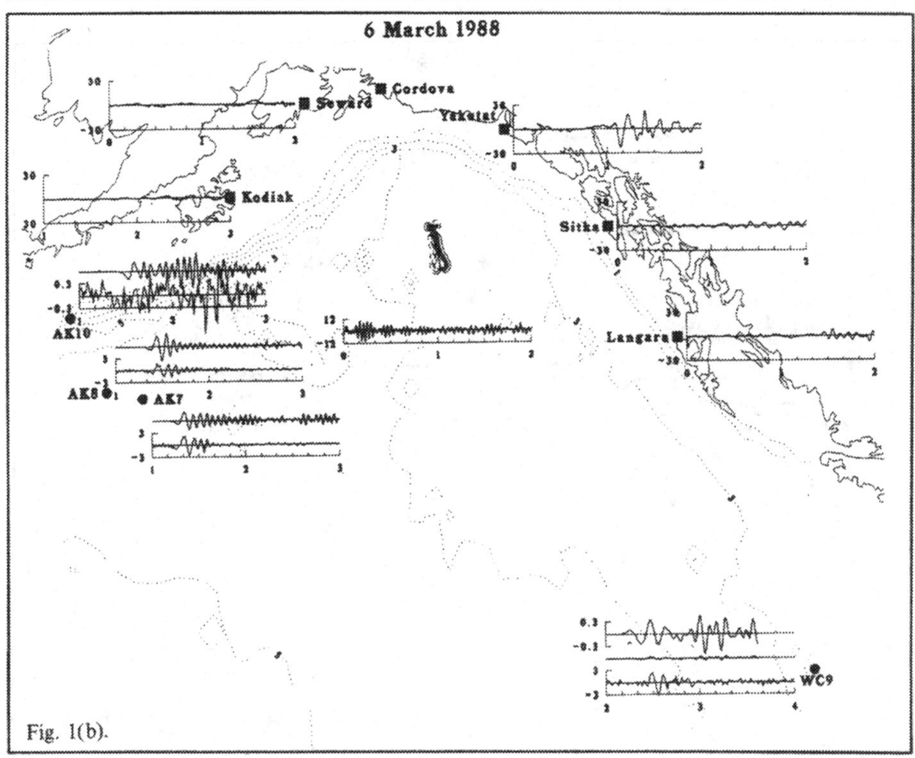
6 March 1988
Cordova
Seward
Yakutat
Kodiak
Sitka
Langara
AK10
AK8
AK7
WC9
Fig. 1(b).

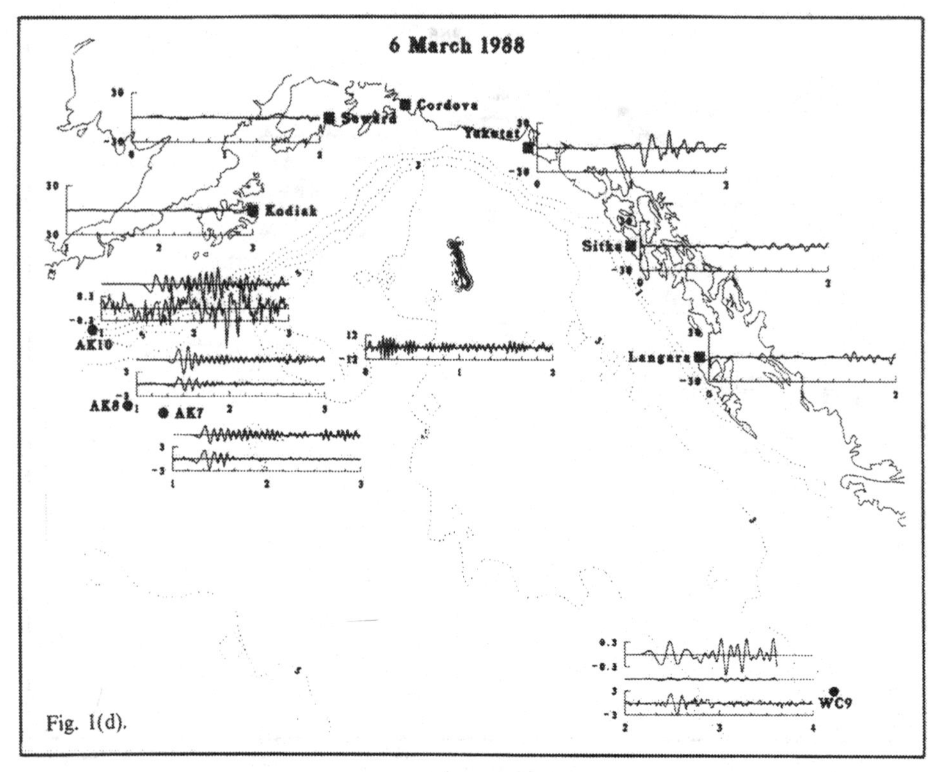
6 March 1988
Cordova
Seward
Yakutat
Kodiak
Sitka
Langara
AK10
AK8
AK7
WC9

Fig. 1(d).

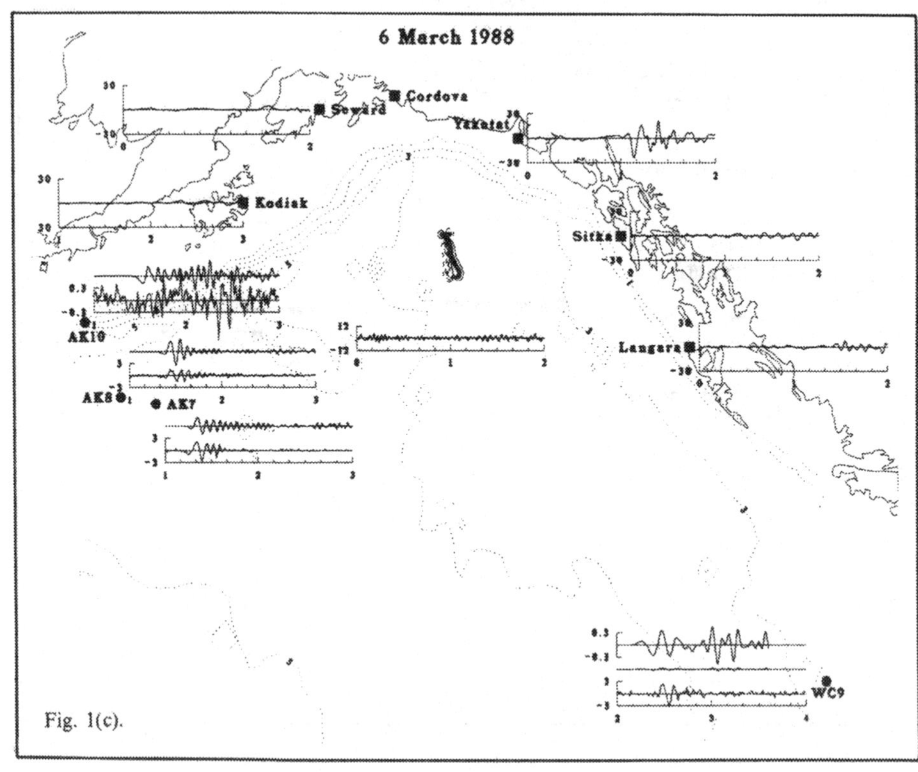
6 March 1988
Cordova
Seward
Yakutat
Kodiak
Sitka
Langara
AK10
AK8
AK7
WC9

Fig. 1(c).

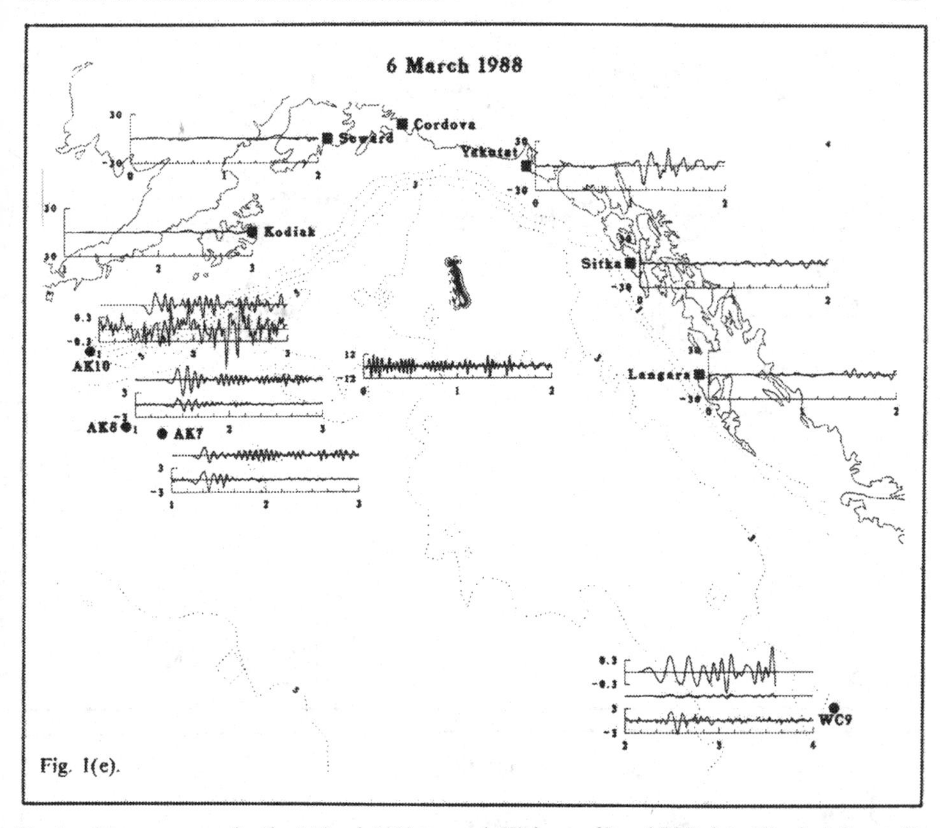

Fig. 1(e).

Fig. 1. Summary maps for the 6 March 1988 tsunami. High-pass filtered BPR data (60 min 3 db cutoff) and SWAN model output are plotted near each BPR station (solid circles), with model output displaced and plotted to the same scale directly above the BPR data; model output (except for the source) has also been shifted in time and scaled in amplitude as explained in the text. Coastal tide gauge data (band-pass filtered: 3 and 60 min 3 db cutoffs) are also shown at each location (squares); stations lacking data plots did not record a tsunami. Horizontal axes are time in hours after the earthquake main shock, and vertical axes are in centimeters. Bathymetric contour labels are in kilometers. Map projection is azimuthal equidistant with pole at the earthquake epicenter, so that great circles are straight lines and distances from the epicenter are true to scale. The quadrupole features southwest of Yakutat are contours of model estimates of u_z, the vertical seafloor displacement, for a total slip in the fault plane of 10 m. The contour interval is 10 cm, and outermost contours have the values ± 10 cm; solid lines are positive displacement (uplift), dashed are negative (subsidence). Simulations were performed with the two different fault plane models and the displacement time histories presented in Table III. (a) First seismic model with zero rise time and infinite rupture velocity. (b) Second seismic model with zero rise time and infinite rupture velocity. (c) Second seismic model with 90 sec rise time, infinite rupture velocity. (d) Second seismic model with zero rise time, 2.2 km/sec rupture velocity. (e) Second seismic model with zero rise time, 0.7 km/sec rupture velocity.

than the abyssal stations, so that depth attenuation of higher frequency pressure signals is less efficient; furthermore, both stations are also in regions of numerous microseisms, which could be an additional source of bottom pressure signals (Bradner, 1962; Takahashi, 1981; Filloux, 1982).

The BPRs which measured the tsunamis are self-contained units that utilize a Paroscientific Digiquartz pressure sensor and a 1.4-Megabyte tape recorder; a

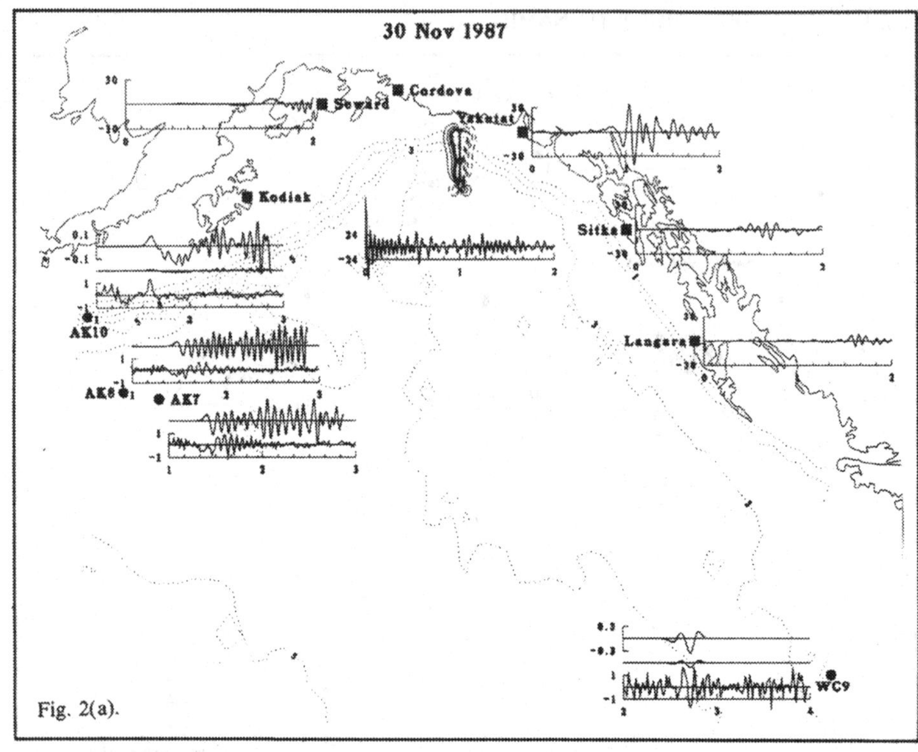
30 Nov 1987
Cordova
Seward
Yakutat
Kodiak
Sitka
Langara
AK10
AK8
AK7
WC9
Fig. 2(a).

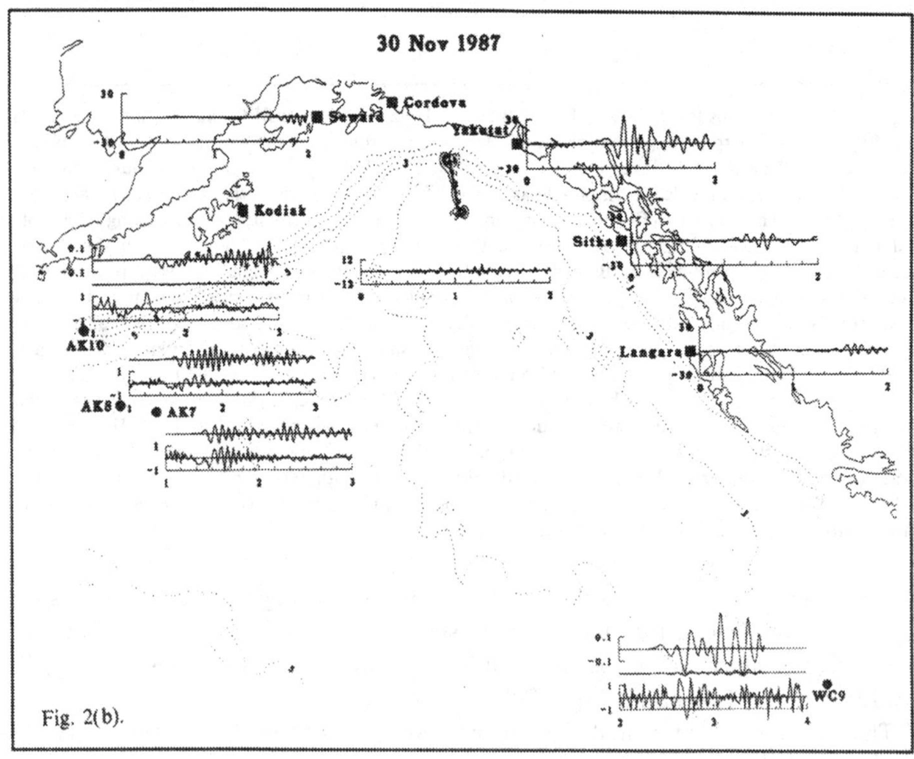
30 Nov 1987
Cordova
Seward
Yakutat
Kodiak
Sitka
Langara
AK10
AK8
AK7
WC9
Fig. 2(b).

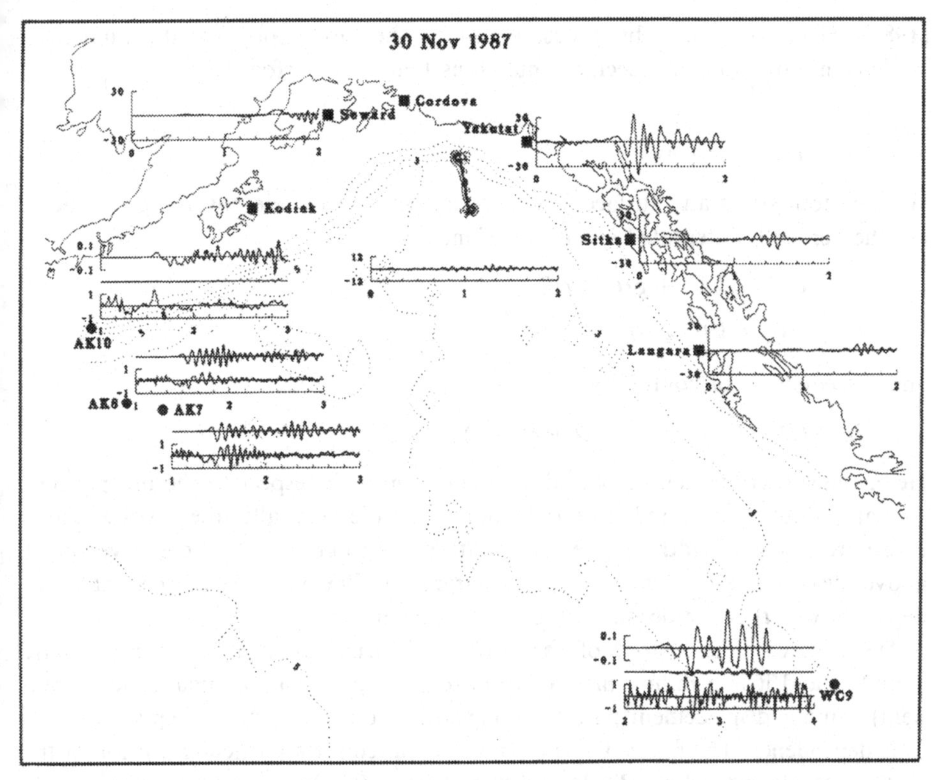

Fig. 2. Summary maps for the 30 November 1987 tsunami. Conventions are the same as for Figure 1. (a) First seismic model with zero rise time, infinite rupture velocity. (b) Second seismic model with zero rise time, infinite rupture velocity. (c) Second seismic model with 90 sec rise time, infinite rupture velocity.

56.25 sec average pressure is recorded every 56.25 sec. A one-millibar (mb) increment of observed pressure corresponds to an equivalent increment of 0.99185 cm of standard seawater, and all BPR observations have been converted to units of cm. The corresponding sensor resolution is then less than 1 mm equivalent seawater head for periods greater than a few minutes. All records were subjected to tide removal and a 60 min high-pass filter to isolate the tsunami frequency band. A more detailed description of the instrumentation and data processing procedures is provided by Eble *et al.* (1989). The coastal tide gauges are standard, gas-purged, pressure recording 'bubblers', characterized by a system accuracy of about 1% full scale in most cases (Young, 1977). The analog records were digitized and also subjected to tide removal and 3–60 min band-pass filtering to isolate tsunami energy.

3. Numerical Simulations

SWAN (not an acronym) is a finite-difference implementation of the nonlinear shallow water equations; a detailed description of this code is provided by Mader

(1988). Here we give a brief description of the model physics, the numerical implementation, and the specific simulations that were performed.

3.1. *Governing Equations*

With bottom stress and surface forcing terms set to zero, the governing equations are the vertically integrated equations of motion

$$U_t + UU_x + VU_y + gH_x = fV,$$

$$V_t + UV_x + VV_y + gH_y = -fU,$$

and the equation of continuity

$$H_t + (D + H - R)_x U + (D + H - R)_y V = R_t.$$

Here, the subscripts denote partial differentiation with respect to the independent spatial variables (x, y) and time t; U and V are the vertically integrated x and y transports per unit width; g is the acceleration due to gravity; H is the wave height above mean sea level; f is the Coriolis parameter; $D(x, y)$ is the permanent water depth; $R(x, y, t)$ is the ocean bottom displacement term.

The detailed time history of the seafloor deformation $R(x, y, t)$ can be quite complex, and it is common practice to introduce $u_z(x, y)$ as the final (and permanent) seafloor displacement pattern, along with two simplifying concepts regarding time dependence. The first is the *rise time*, τ, required for a particular portion of the seafloor to reach the final displacement pattern, $u_z(x, y)$; the second is the *rupture velocity*, v_r, at which the initial onset of seafloor deformation propagates along the fault axis. We used Okada's (1985) expressions for permanent surface deformation due to motion in an inclined, rectangular fault plane to estimate $u_z(x, y)$, and investigated two particularly simple categories of seafloor motion time history. The first category assumed infinite propagation velocity, but finite rise time in the form of a linear ramp, so that

$$R(x, y, t) = \begin{cases} (t/\tau)u_z(x, y), & \text{for } 0 \leqslant t \leqslant \tau, \\ u_z(x, y), & \text{for } \tau < t. \end{cases}$$

The second category assumed a finite propagation velocity and zero rise time, so that

$$R(x, y, t) = \begin{cases} 0, & \text{for } 0 \leqslant t \leqslant r/v_r, \\ u_z(x, y), & \text{for } r/v_r < t, \end{cases}$$

where r is the along-axis distance from the epicenter.

Reflecting and continuum boundary conditions, as described by Mader (1988), were applied at the coastline and open ocean boundary, respectively. A series of test runs were performed using 5, 10, and 15 sec time steps, and no significant differences were found in the output; a 10 sec time step was subsequently used for all runs presented here.

3.2. *Grid Domain and Output Stations*

Bathymetric data were available on both a $5' \times 5'$ grid ($1'$ is equal to one-sixtieth of a degree) provided by NOAA's Geophysical Data Center and a $10' \times 7.5'$ grid utilized by Kowalik and Murty (1987) in a previous study. A test run with each grid was performed, and no significant differences were found in the results; the $10' \times 7.5'$ grid was therefore employed in our computations. The mesh consisted of 347 nodes in the x direction (positive to the east), and 131 nodes in the y direction (positive to the north), such that the computational domain extended from 122° 20′ to 180° West and from 45° to 61° 15′ North; the north-south inter-nodal distance was $dy = 13.9$ km, and the east–west increments varied from $dx = 13.1$ km at 45° latitude to $dx = 8.9$ km at 61° 15′ N. The grid domain, BPR locations, model output locations (hydro stations) and bathymetric contours are presented in Figure 3, and the location and local water depth of each BPR station are provided in Table I.

In designating a particular nodal point as a SWAN hydro station, some tradeoff was required in attempting to minimize two factors – distance to the BPR position, and the difference in water depth at the hydro station and the BPR. In the case of abyssal stations AK7 and AK8, bathymetric gradients are very small and the corresponding differences are only about 15 km in position and 10 m in depth. However, significantly larger depth gradients characterize the local bathymetry at

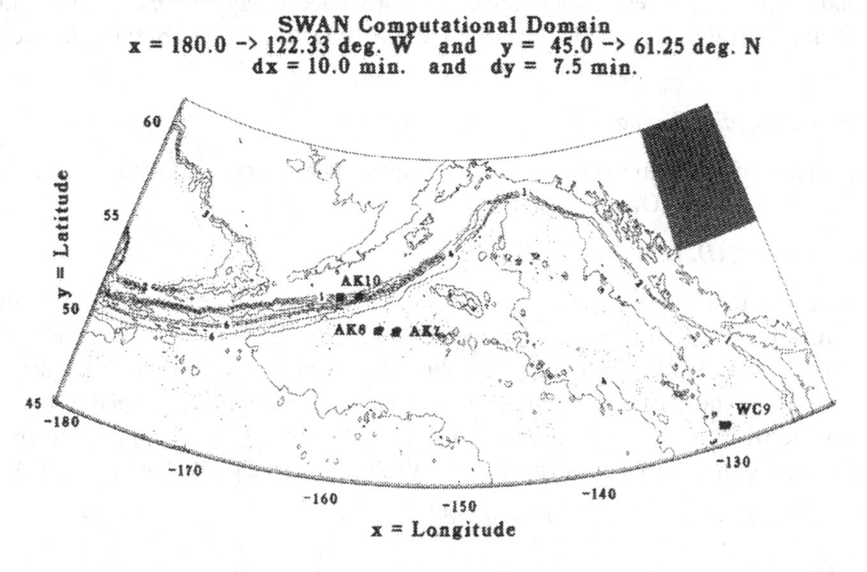

Fig. 3. Reference map of the SWAN model domain. Also shown are the BPR locations (solid circles) and the corresponding grid node hydrostations (solid squares) for which model output was retained for comparison with the observations. The bathymetric contours shown here were derived from data on a $10' \times 10'$ grid ($1'$ is equal to one-sixtieth of a degree), somewhat coarser than the $10' \times 7.5'$ bathymetric data set employed for the numerical simulations; the contour interval is 500 m and the labels are in kilometers. The size of individual computational cells and the overall density of the $10' \times 7.5'$ grid mesh are illustrated in the upper-right-hand corner of the map.

Table I. BPR station information

Station	Latitude (° N)	Longitude (° W)	Depth (meters)
AK7	52.73	155.00	4463
AK8	52.73	156.47	4535
WC9	45.97	129.99	1527
AK10	54.28	158.51	1656

stations WC9 on the mid-ocean ridge and AK10 on the trench slope. As a consequence, one grid node was only 18 km distant but 486 m deeper than the known position and depth of station WC9, while a neighboring node was 26 km distant but only 17 m deeper. Similarly, the depth difference was over 1800 m at the grid node nearest the AK10 position, while a cell 109 km distant represented a depth difference of 126 m. Test simulations revealed only minor differences in model output at candidate grid cells; we therefore chose the cells with the more representative depth values.

At hydro stations AK7, AK8, and AK10, the model output is free of spurious open boundary signals for approximately four hours after tsunami arrival, corresponding to round-trip travel time to the nearest open boundary over a distance of 1300–1600 km. However, some open boundary effects may be present 30 min after tsunami arrival at station WC9, located about 100 km from the southern boundary.

3.3. *Earthquake Simulations*

Estimates of the static vertical surface displacement were computed from expressions of the form (Okada, 1985)

$$u_z = Uf(D, L, W, \phi, \delta, \gamma; r),$$

where U is total slip in the fault plane, and a particular fault plane model is defined by function f in terms of the parameters D, L, W, ϕ, δ, and γ – the fault plane depth, length, width, strike, dip, and slip rake, respectively. The last parameter, r, is related to the elastic properties of the particular medium; for an elastic medium characterized by Lamé's constants λ and μ, and density ρ, the propagation speeds of P-waves and S-waves are given, respectively, by (see, e.g., Bullen and Bolt, 1985)

$$\alpha = [(\lambda + 2\mu)/\rho]^{1/2}, \qquad \beta = (\mu/\rho)^{1/2}$$

so that

$$r \equiv \mu/(\lambda + \mu) = \beta^2/(\alpha^2 - \beta^2) = [(\alpha/\beta)^2 - 1]^{-1}.$$

We set $r = 0.3$ in our simulations; this corresponds, e.g. to the case $\alpha \approx 8.1$ km/s and $\beta \approx 3.8$ km/s, so that $(\alpha/\beta) \approx 2.1$, which is consistent with generally accepted values for Pacific oceanic crust (Brune, 1979; Saito and Takeuchi, 1966).

Since vertical seafloor displacement in this model is directly proportional to the total slip U, the specification of this parameter is somewhat arbitrary. In addition, the nonlinear terms in the governing equations (Section 3.1) are negligible for reasonable seafloor displacements and tsunami wave heights. An important consequence of these two facts is the essential linearity of this problem, so that scaling of solutions is valid; i.e., multiplication by the scale factor s of a tsunami solution field obtained for a particular slip value U_0 is equivalent to the solution field which would have been obtained for total slip equal to $U = sU_0$. The value of U_0 was chosen to be 10 m for all simulations; in addition to scaling conveniently, this value was found to produce maximum seafloor displacement values on the order of a meter and far-field tsunami amplitudes that were of the same order of magnitude as the observations, i.e. a few centimeters.

Table II presents three fault plane model solutions for each earthquake computed shortly after their occurrence (USGS, 1987a, b). In each case, there are significant differences in parameter values obtained by the three methods. Our first simulation of each earthquake used fault plane parameters which more closely approximated those published by Lahr *et al.* (1988). Subsequent simulations of each earthquake used unpublished values which were obtained more recently (Page, 1989). For convenience, we will hereafter refer to these as the *first* and *second* models, or as models 1 and 2.

The first and second simulations of each earthquake and tsunami were performed with time history parameters corresponding to an instantaneous displacement ($\tau = 0, v_r = \infty$). Noninstantaneous displacements were modeled only with the second fault plane models; three cases were simulated for 6 March 1988 and one for 30 November 1987. In the case of the two 6 March simulations with finite rupture velocity, the seafloor deformation was initiated at the southern end of the fault and then allowed to propagate north; this corresponds to the observed location of the epicenter at the southern end of the aftershock zone as described by Lahr *et al.* (1988).

Specific fault plane and time history parameter values for each simulation are listed in Table III. The corresponding model results for each simulation are presented in Figures 1 and 2; also shown are the hydro station time series near the epicenter and the quadrupole patterns located southwest of Yakutat which represent the final seafloor displacement patterns.

3.4. *Scaling of Output*

To facilitate model/BPR comparisons, we chose station AK7 as a reference station and defined a time shift and amplitude scaling factor for the model output which would force agreement with the observations in the magnitude and time of arrival of the first local maxima in the waveforms.

Initially, we sought to define the time shift, Δt, in terms of the leading edge of the wave, which, because the local wavelength and period are theoretically infinite there

Table II. Summary of earthquake parameter estimates. The *Mod* column indicates the type of model used to estimate parameters; 'F' refers to fault plane solutions derived from initial P-wave polarities. 'M' refers to the USGS moment tensor solution, and 'C' refers to the Harvard centroid solution. ϕ, δ and γ are the fault plane strike, dip, and slip rake angles, respectively. D is the hypocentral depth, and the entry 'f' indicates that D was fixed, rather than computed. L and W are the fault length and width. M_0 is the moment estimate. U, the total slip in the fault plane, is inferred from equation (5) with L set to the estimates of Lahr *et al.*, 1988.

Reference	Mod	Date	Time (GMT)	Lat. (°W)	Long. (°N)	ϕ (°)	δ (°)	γ (°)	D (km)	L (km)	W (km)	M_s	m_b	M_0 (10^{20}N-m)	U (m)
USGS, 1987b	F	6 Mar 88	223538.14	56.953	143.032	151	88	−169	10f		20.0	7.6	6.8		
USGS, 1987b	M	6 Mar 88				338	87	133	18		36.0			1.0	0.6
USGS, 1987b	C	6 Mar 88	223551.6	57.37	143.53	182	75	−168	15f		31.1			4.9	3.3
Lahr *et al.*, 1988	F	6 Mar 88	223536.38	57.23	142.78	175	69	−178	10f	110	21.4	7.6	6.8	4.9	4.8
Page, 1989	F	6 Mar 88	223536.41	56.856	143.082	349	88	172	10f		20.0			4.9	5.2
USGS, 1987a	F	30 Nov 88	192319.59	58.679	142.786	352	90	183	10f		20.0	7.6	6.7		
USGS, 1987a	M	30 Nov 88				176	88	−179	20		40.0			6.6	2.7
USGS, 1987a	C	30 Nov 88	192340.3	58.17	142.04	355	70	−172	15f		31.9			7.3	3.7
Lahr *et al.*, 1988	F	30 Nov 88	192316.39	58.91	142.76	171	90	166	10f	140	20.0	7.6	6.7	7.3	6.1
Page, 1989	F	30 Nov 88	192315.10	58.634	142.766	350	88	−177	10f		20.0			7.3	6.1
USGS, 1987a	F	17 Nov 88	084653.32	58.586	143.270	274	83	2	10f		20.2	6.9	6.6		
USGS, 1987a	M	17 Nov 88				273	89	−8	18		36.0			0.63	1.0
USGS, 1987a	C	17 Nov 88	084704.5	58.87	143.62	262	57	−6	15f		35.8			0.66	1.1
Lahr *et al.*, 1988	F	17 Nov 88	084650.89	58.80	143.11	275	83	2	10f	40	20.2	6.9	6.6	0.66	1.9

Table III. Numerical simulation summary. For convenience, a reference figure is given in the first column, and the model number in the second column is used in the text to refer to a specific set of fault plane parameters. Time history parameters τ and v_r are the rise time and rupture velocity of the seafloor displacement, respectively. The time-shift, Δt, and total slip in the fault plane, U, were chosen to force agreement of the model results with observations at station AK7, with positive Δt indicating early arrival of the model tsunami. Maximum, minimum, and rms values of the vertical displacement u_z, and the areal coverage A of the displacement pattern refer to that region for which $|u_z| \leqslant 0.01|u_z|_{max}$. Earthquake moment, M_0, was estimated using Equation (4).

Ref. Fig.	Mod	Date	Lat. (°W)	Long. (°N)	ϕ (°)	δ (°)	γ (°)	D (km)	L (km)	W (km)	τ (s)	v_r (km/s)	Δt (min)	Max u_z (cm)	Min u_z (cm)	rms u_z (cm)	A (10^3 km^2)	M_0 (10^{20} N-m)	$U = 10\zeta/H$ (m)
1(a)	1	6 Mar 88	57.125	143.000	180	69	−178	12.6	110	25.0	0	∞	5.4	54.7	−60.1	7.9	21.6	4.3	3.6
1(b)	2	6 Mar 88	56.750	143.333	349	88	172	10	110	20.0	0	∞	8.2	59.0	−50.1	10.0	24.0	4.7	5.0
1(c)	2	6 Mar 88	56.750	143.333	349	88	172	10	110	20.0	100	∞	7.8	63.7	−54.5	10.8	24.0	5.1	5.4
1(d)	2	6 Mar 88	56.750	143.333	349	88	172	10	110	20.0	0	2.20	8.3	62.5	−53.5	10.6	24.0	5.0	5.3
1(e)	2	6 Mar 88	56.750	143.333	349	88	172	10	110	20.0	0	0.73	7.6	73.2	−62.6	12.4	24.0	5.9	6.2
2(a)	1	30 Nov 88	58.750	143.000	180	90	166	12.6	140	25.0	0	∞	5.4	67.2	−67.7	15.8	32.2	7.2	4.8
2(b)	2	30 Nov 88	58.750	143.000	350	88	−177	10	140	20.0	0	∞	11.7	61.2	−78.2	10.9	23.9	8.2	6.8
2(c)	2	30 Nov 88	58.750	143.000	350	88	−177	10	140	20.0	100	∞	10.8	72.9	−93.2	13.0	23.9	9.8	8.1

(Kajiura, 1963), propagates at the long wave celerity

$$c_g \approx (gD)^{1/2}. \tag{1}$$

However, the leading edge is also vanishingly small by definition, and its temporal growth as seen at a particular observation point is a complex function of the source and propagation path characteristics. For model results, specification of the leading edge is simple – a matter of locating the time of the first nonzero value for H; this initial excursion can be exceedingly small, but easily discernible because the model output is free of background noise. Observational data, however, are characterized by nonzero background energy levels; as a consequence, the unambigious specification of leading edge arrival time is very subjective and prone to considerable error. We therefore defined Δt as the arrival time of the first local maximum in the packet, rather than the leading edge; in addition to reducing ambiguity, this has the advantage of dealing with that portion of the wave which is of most practical interest in an operational sense – i.e., the maximum of the first tsunami wave.

We next computed the scaling factor $s = \zeta/H$ which would force agreement at station AK7 in the amplitude of the first observed tsunami wave maxima, ζ, and the first computed model wave maxima, H. SWAN records at the remaining hydro stations (not the source) were subsequently shifted in time by the amount Δt and multiplied by the scaling factor s. This procedure helped to clarify waveform similarities and differences in a systematic way. The resulting time-shift and the total slip, $U = sU_o = 10\zeta/H$, are listed in Table III for each of the simulations. Also listed are the extrema and rms values of u_z, and the areal extent, A, of the displacement pattern over which $|u_z| \leqslant 0.01|u_z|_{max}$.

4. Results

At abyssal stations AK7 and AK8, the simulations for 6 March were in better agreement with the observed tsunami wave packets than the simulations of 30 November. For both tsunamis, one or more simulations produced a wave packet that was roughly comparable to the observed waveform in terms of the packet length and number of individual waves. However, the observed waveform on 30 November is the more complex, with a very low frequency negative displacement at the leading edge that is not apparent in the 6 March waveform; it is this feature that was not reproduced by the 30 November simulations (Figures 1 and 2).

Since better agreement was found in the case of the 6 March tsunami, and since a more varied set of five simulations was performed for this tsunami, those results will be discussed first. Also, since there is a substantial amount of information to examine, we will organize the comparisons by discussing each station in turn, focussing on certain specific features of the tsunami waveforms – first motion, arrival time, position and amplitude of the maximum wave, the periods of individual waves, and the wave packet shape and duration. For convenience, we will also refer to individual model runs by the corresponding number of the figure that displays the data.

4.1. *The 6 March 1988 Tsunami*

There are substantial differences in the seafloor displacement patterns corresponding to the first and second fault plane models; the first is asymmetric about the north-south strike axis and very nearly antisymmetric in the perpendicular east-west direction (Figure 1a), while the opposite is true of the second (Figure 1b). This is a consequence of a rake angle near $\pm 180°$ for the first model and a dip angle near $\pm 90°$ for the second (Table III). Although the two deformation patterns were very similar in terms of areal extent and maximum, minimum, and rms vertical displacement (Table III), the tsunami model produced substantially different results for each seismic model; in part, this must be due to significant differences in the tsunami directivity of each pattern.

At AK7, the reference station, first motion was negative for all five simulations; this is in agreement with the data and consistent with negative seafloor displacement in the southwest quadrant of both models. Tsunami arrival times were early for all simulations – 5.4 min in the case of the first seismic model, and about 8 min for the second. In particular, the arrival times for runs 1(a) and 1(b), with identical time histories but different seismic models, differed by 3.8 min (Table III). This difference is due primarily to the position and orientation of the two different seafloor deformation patterns. The great circle distances from AK7 to the nearest edge of the first and second deformation patterns are 840 and 870 km, respectively – a difference of 30 km. Since the average is 855 km and the propagation time for the first peak is 81 min, then the apparent propagation speed is about 10.5 km/min. The expected difference in arrival times is therefore approximately $30/10.5 \approx 2.9$ min. This is close to the difference observed, and the 1 min BPR sampling rate may account for an additional 30 sec of the discrepancy.

The maximum wave height of simulation 1(a) is also about 39% larger than that for 1(b), as indicated by the ratio of their respective scaling factors ζ/H (Table III). This difference is too large to be accounted for by the effects of radial spreading over the additional 30 km. For an r^{-1} decay law, the effect on the wave height ratio at the two stations would be $870/840 \approx 1.04$, or only about 4%. Refraction, source-station geometry, and source directivity must therefore play important roles.

For the first model run, 1(a), the length of the wave packet was 11 min shorter than the observed length of 33 min, and the individual wave periods were too long; in contrast, the individual wave periods were in good agreement for the second model, but the length of the model wave packet was 7 min longer than observed.

Changes in the source time history of the second model had little effect on either the arrival time or the amplitude of the first positive wave. This is clear from the near constant values for Δt and ζ/H seen in Table III for runs 1(b)–1(e); in each of the four simulations the model tsunami arrived at AK7 about 8 min earlier than actually observed ($\Delta t \approx 8$), and the first positive wave height of the model was about twice that observed ($\zeta/H \approx 0.5$). There is, in fact, good agreement with the observed periods and amplitudes of the individual waves for the first few cycles of the observed waveform for all four runs.

The primary difference between runs is not apparent for several cycles, after which it becomes evident that different time histories produce differences in the low-frequency amplitude modulation. This is to be expected, since, generally speaking, amplitude modulation arises from constructive and destructive interference of wave energy propagating from different regions of a finite source, and the space-time histories of these sources are all different. Simulation 1(e) appears to agree most closely with the data, producing an initial wave packet with a duration which most nearly matches that of the observations. However, we note that the same simulation subsequently produces additional wave packets not evident in the data. In fact, this discrepancy is common to all five simulations; i.e., all are characterized by one or more additional wave packets (albeit less energetic), in contrast to the single packet observed (Figures 1a–1e, Sta. AK7).

At AK8, the time and amplitude of first maximum are in fair agreement for all five runs once Δt and ζ/H (computed at AK7) are applied. Some aliasing of the BPR data is apparent at AK8; the first few tsunami cycles have a mean period of about 5 min, and 1 min BPR measurements somewhat under-sample the waveform. In spite of the aliasing, the data do suggest that the first maximum is the largest at AK8 (as it is at AK7), in agreement with the first seismic model (Figure 1a). In apparent disagreement the second wave is the largest for each simulation employing the second seismic model, regardless of the particular time history assumed; however, packet length and individual wave periods are in better agreement (Figures 1b–1e).

In this regard, it is notable that the duration of the first computed wave packet at AK8 is well-defined and nearly constant (≈ 30 min), regardless of changes in the source time history for each run; this is in direct contrast to the results at AK7, in which the computed amplitude modulation varies significantly with the source time history. This difference in results for two stations separated by only 100 km is important as an illustration of the sensitivity of tsunami data/model comparisons to both station location and fundamental assumptions regarding the model source.

At WC9, the BPR data display two clear cycles of a tsunami waveform, with amplitudes on the same order (≈ 3 cm) as the abyssal stations, AK7 and AK8; the characteristic low frequency negative displacement observed at AK7 and AK8 is somewhat obscured by higher frequency energy, but first motion does appear to be negative. SWAN output is strikingly different for each seismic model. The first motion in 1(a) is negative, in agreement with the data; however the first maximum arrives about 6 min early (for a total of 11.4 min, if the 5.4 min shift is taken into account), and the amplitude of the first wave is too small by a factor of three. There are even larger discrepancies for the second seismic model runs, 1(b–e). The initial tsunami motion is positive, and the period of the first wave is too long; in addition, the amplitudes are too low by an order of magnitude, so that additional graphs with a much smaller vertical scale were needed in Figures 1(b–e).

At AK10, the data are somewhat questionable because of the much reduced signal-to-noise ratio (SNR) (Figures 1a–1e, Sta. AK10). However, the data are suggestive of the first two cycles of a very small amplitude (≈ 0.3 cm) tsunami; the first maximum arrives at about the arrival time indicated by the model (shifted by Δt), and the period of the first wave is similar to that observed at AK7 and AK8. As in the case of the model output for AK8, it can be seen that the first few cycles of the tsunami at AK10 were relatively insensitive to the source time history and changed very little from one simulation to the other. Several prominent, low-frequency oscillations can also be seen in the record, commencing about 2 h 15 min after the main shock; these may be local shelf or trench modes excited by the incident tsunami.

4.2. *The 30 November 1987 Tsunami*

Generally poor agreement was found between the BPR observations and model simulations for the two different seismic models. The displacement patterns for each model were significantly different, both in shape and individual displacement statistics. The first model is characterized by a much more prominent northern dipole than the second (Figure 2). The first model is also the most energetic of the two, characterized by a 21 to 45% greater rms displacement over a 35% larger area (Table III). As a consequence, the computed tsunami waveforms were quite different for run 2(a) and runs 2(b and c).

At AK7 and AK8, the first maximum is 40% larger for the first model than the second (Table III), consistent with the more energetic nature of the first model. First motion for all simulations is positive, in clear disagreement with the very-low-frequency negative displacement observed at these two sites. A clearly defined wave packet is not evident in the waveforms for run 2(a); in fact, model wave energy is seen to increase at about the time observations indicate termination of the wave packet. The second seismic model does produce finite wave packets, but the envelope shapes and the numbers and periods of individual waves are in generally poor agreement (Figures 2b and 2c). As for the 6 March simulations, additional wave packets are also generated which are not observed in the data.

At WC9, first motion is negative for the first seismic model and positive for the second. Unfortunately, the observed first motion cannot be determined due to the very low SNR of the data. Both models produce amplitudes that are much too low; however, there is a surprising correspondence between the computed and observed time of positive and negative extrema for the first two waves, especially in the case of runs 2(b) and 2(c).

At AK10, all three simulations display another very interesting correspondence of the first computed maxima with a prominent positive displacement in the data.

This strongly suggests that the observed feature is, in fact, the tsunami wave form, albeit severely distorted through interaction with the complex local bathymetry.

5. Discussion

The relative magnitudes of the tsunami waveforms were reproduced by the model, with the exception of station WC9 on the midocean ridge. At all stations and in all simulations, the model tsunami arrived somewhat early. AT AK7, the discrepancy was 5–8 min on 6 March, and 5–12 min on 30 November (Table III). About 2 minutes can be accounted for by the fact that, away from the leading edge, the local group velocity is somewhat less than that which is implicit in the long wave formalism of the model, i.e., Equation (1). Thus, the group speed is given by the well-known expression

$$c_g = (\omega/2k)[1 + 2kD/\sinh(2kD)], \tag{2}$$

where the wavenumber, k, is related to the radian frequency, ω, by the dispersion relation

$$\omega^2 = gk\tanh(kD). \tag{3}$$

The first maxima of the observed tsunami waveforms are characterized by local periods on the order of 10 minutes. For mean depths of 4000 and 5000 m, this expression yields values which are 2.2 and 2.8% smaller than the long wave approximation expressed by Equation (1), so that this effect would account for only 2 min or so at AK7. This leaves a 3 to 6 min discrepancy for the 6 March tsunami and 3 to 9 min difference for the 30 November case.

Travel-time discrepancies may also be associated with errors in the distance between a hydrostation and the nearest edge of the model source; this error is a function of two factors – differences in the actual and computed deformation pattern and differences in the geographical position of these patterns. The first and second estimates of the epicenter location differ by 45 km for 6 March and 31 km for 30 November. These distances correspond to 3 or 4 min of travel time at tsunami propagation speeds of 10–12 km/min. An additional 3 or 4 min error could easily be introduced by differences in the actual and computed deformation; since the length scale of a lobe in the computed quadrupole pattern is tens of kilometers, for example, a missing or additional lobe would introduce errors of this order. Finally, the finite depth grid is only an approximation to the real bathymetry, and the general tendency of unresolved depth variations is to increase computed wave speeds, as discussed by Danard and Murty (1989).

Most simulations were also characterized by an apparent excess of tsunami energy at stations AK7 and AK8 that continued to arrive long after the observed tsunami amplitude had died out. Open boundaries are too distant to be responsible, so test runs were performed in which a variable bottom friction term with De Chezy's coefficient was introduced and in which the percentage of reflected energy from the coastal boundaries was varied via a transmission coefficient. Friction had to be increased to an unrealistic level before the model output was significantly affected at the low frequencies of interest; also, most energy appeared to be reflected by offshore bathymetric variations, rather than by coastal boundaries, so that variations in the transmission coefficient also had a negligible effect. We therefore speculate that this discrepancy was caused by the fundamental problem of accurately specifying and modeling the seafloor displacement corresponding to the tsunami source function.

Of all the simulations, perhaps 1(e) displayed the best agreement with the observations at the abyssal stations, AK7 and AK8, in terms of first motion, amplitude and periods of individual waves, and wave packet length; there are also certain similarities in the computed and observed waveforms for the first one or two cycles at stations AK10 and WC9, although the amplitude at WC9 is much too low. To the degree that these results imply a modicum of similarity between the actual and computed source mechanisms, the values for Δt and U in Table III suggest that the 6 March earthquake may have been characterized by a rather slow rupture velocity in the order of 0.7 km/sec, and that the total slip in the fault plane was about 6 m. For the assumed rake angle of 172°, effectively all of the slip motion would be along-strike. This is somewhat high when compared with estimates reported by Lisowski and Savage (1989); they found total right-lateral slip due to *both* the 30 November and 6 March earthquakes to be 2.9 ± 1.2 m using geodetic means, and 4.5 m using the seismic moment.

For a rectangular fault plane model, we have the relationship (e.g., Bullen and Bolt, 1985)

$$M_0 = \mu LWU \tag{4}$$

or

$$U = M_0/(\mu LW). \tag{5}$$

The first expression was used to compute estimates of M_0 presented in Table III, using estimates of U derived from the scaling of each simulation; the value $\mu = \rho\beta^2 = 4.3 \times 10^{11}$ dynes/cm^2 was used, corresponding to $\rho = 3.0$ gm/cm^3 and $\beta = 3.8$ km/s (Saito and Takeuchi, 1966). Similarly, Equation (5) and the estimates of M_0, L, and W presented in Table II were used to compute the values of U entered in the last column. These computations yield a slip of 4.8–5.2 m (Table II) for the fault plane parameters estimated by Lahr *et al.* (1988) and Page (1989). Furthermore, the slip of 5.3 m suggested by simulation 1(d) produces the estimate $M_0 = 5.0 \times 10^{20}$ N-m (Table III), which is within a few percent of the value 4.9×10^{20} N-m obtained via the Harvard centroid solution (Table II).

6. Summary and Conclusions

A series of long-wave tsunami numerical simulations have been performed for the 30 November 1987 and 6 March 1988 Alaskan Bight earthquakes. Two different fault plane seismic models were used for each tsunami: a set published by Lahr *et. al* (1988) and a second, unpublished, set computed more recently.

The model reproduced the relative magnitudes of the observed tsunamis at three of the four stations; the exception was station WC9 in the Axial Caldera of the Juan de Fuca midocean ridge. Fair agreement was also obtained for the first few waves in the packet of the 6 March tsunami, using the most recent seismic model; the best pair of simulations were characterized by rupture velocities of 0.7 and 2.2 km/sec and total slip in the fault plane of 5.3 and 6.2 m. These values for the slip, and the corresponding estimates of the seismic moment, are in good agreement with estimates based on the Harvard centroid solution for the 6 March earthquake. This is an encouraging result, which implies that the modeled source characteristics bear at least a qualitative resemblance to the actual seafloor displacement history. The waveform of the 30 November 1987 tsunami exhibited very-low-frequency negative displacement of the leading wave which was not reproduced by the numerical model with the sources studied. For both tsunamis, the calculated tsunami arrivals were early, and the model produced a small second tsunami packet that was not observed.

It is clear that the model/data comparison results are quite sensitive to the specific source assumed and, since even the simplest fault plane model requires the specifications of at least eight independent parameters, the systematic exploration of candidate sets of fault plane parameter can be expensive and time-consuming.

Furthermore, the fault plane formalism may be inadequate to specify features important to the tsunami simulation. Additional simulations of the type presented here are best attempted only after more sophisticated analyses of the seismic data base provide an improved understanding of the seafloor deformation mechanism.

Alternatively, the tsunami waveforms themselves could be used to help constrain the fault mechanisms, using geophysical inversion techniques such as those recently employed by Satake (1989). The potential disadvantage of this approach, however, would be the elimination of the tsunami data set as an important independent check on seismically derived estimates.

Acknowledgements

Dr R. Page of the U.S. Geological Survey kindly provided us with the most recent estimates of the earthquake fault plane parameters. Professor Z. Kowalik of the University of Alaska Institute of Marine Sciences provided the bathymetric grid used in the numerical computations. This work is contribution No. 1194 of the Pacific Marine Environmental Laboratory's Tsunami Project.

References

Bernard, E. N. and Goulet, R.: 1981, *Tsunami Research Opportunities*, National Science Foundation, Washington, D.C.

Bernard, E. N. and Milburn, H. B.: 1985, Long-wave observations near the Galapagos Islands, *J. Geophys. Res.* **90**, 3361–3366.
Bradner, H.: 1962, Pressure variations accompanying a plane wave propagated along the ocean bottom, *J. Geophys. Res.* **67**, 3631–3633.
Braddock, R. D.: 1969, On tsunami propagation, *J. Geophys. Res.* **74**, 1952–1957.
Brune, J. N.: 1979, Surface waves and crustal structure, in P. J. Hart, (ed.), *The Earth's Crust and Upper Mantle*, Geophys. Mongr., No. 13, Amer. Geophys. Union, pp. 230–242.
Bullen, K. E. and Bolt, B. A.: 1985, *An Introduction to the Theory of Seismology*, 4th edn, Cambridge Univ. Press.
Danard, M. B. and Murty, T. S.: 1989, On sources of error in calculation of tsunami travel times, *Sci. Tsunami Hazards*, **7**, 73–78.
Eble, M. C., Gonzalez, F. I., Mattens, D. M., and Milburn, H. B.: 1989, Instrumentation, field operations, and data processing for PMEL deep ocean bottom pressure measurements, NOAA Tech. Memo ERL PMEL-89.
Filloux, J. H.: 1982, Tsunami recorded on the open ocean floor, *Geophys. Res. Lett.* **9**, 25–28.
Gonzalez, F. I., Bernard, E. N., and Milburn, H. B.: 1987a, A program to acquire deep ocean tsunami measurements in the north Pacific, *Proc. Coastal Zone '87*, pp. 3373–3381.
Gonzalez, F. I., Bernard, E. N., Milburn, H. B., Castel, D., Thomas, J., and Hemsley, J. M.: 1987b, The Pacific tsunami observation program (PacTOP), *Proc. 1987 Inter. Tsunami Symp., IUGG*, pp. 3–19.
Kowalik, Z. and Murty, T. S.: 1987, Influence of the size, shape and orientation of the earthquake source area in the Shumagin Seismic gap on the resulting tsunami, *J. Phys. Ocean.* **17**, 1057–1062.
Lahr, J. C., Page, R. A., Stephens, C. D., and Christensen, D. H.: 1988, Unusual earthquakes in the Gulf of Alaska and fragmentation of the Pacific plate, *Geophys. Res. Lett.* **15**, 1483–1486.
Lisowski, M. and Savage, J. C.: 1989, Deformation in the Yakataga Seismic gap, Southern Alaska, associated with the Gulf of Alaska earthquakes of November 1987 and March 1988, (Abstract) *EOS Trans.* **44**, 1439.
Page, R. A.: 1989, Personal communication.
Kajiura, K.: 1963, The leading wave of a tsunami, *Bull. Earthquake Res. Inst. Univ. Tokyo* **41**, 525–571.
Mader, C. L.: 1988, *Numerical Modeling of Water Waves*, University of California Press, Los Angeles, California.
Mei, C. C.: 1983, *The Applied Dynamics of Ocean Surface Waves*, Wiley, New York, Chapter 2.
Okada, Y.: 1985, Surface deformation due to shear and tensile faults in a half-space, *Bull. Seism. Soc. Am.* **75**, 1135–1154.
Raichlen, F.: 1985, Report of Tsunami Research Planning Group, National Science Foundation, Washington, D.C.
Saito, M. and Takeuchi, H.: 1966, Surface waves across the Pacific, *Bull. Seism. Soc. Am.* **56**, 1067–1091.
Satake, K.: 1989, Inversion of tsunami waveforms for the estimation of heterogeneous fault motion of large submarine earthquakes: the 1968 Tokachi-oki and 1983 Japan Sea earthquakes, *J. Geophys. Res.* **94**, 5627–5636.
Takahashi, M.: 1981, Telemetry bottom pressure observation system at a depth of 2200 meter, *J. Phys. Earth* **29**, 77–88.
USGS: 1987a, Preliminary determination of epicenters, monthly listing, U.S. Dept. of the Interior/Geological Survey, November.
USGS: 1987b, Preliminary determination of epicenters, monthly listing, U.S. Dept. of the Interior/Geological Survey, March.
Young, S. A.: 1977, Users guide for the gas-purged pressure recording (Bubbler) tide gage, *National Ocean Survey*.

Natural Hazards **4**: 141–159, 1991.

Investigation of Long Waves in the Tsunami Frequency Band on the Southwestern Shelf of Kamchatka

P. D. KOVALEV, A. B. RABINOVICH and G. V. SHEVCHENKO
U.S.S.R. Academy of Sciences, Far East Branch, Institute of Marine Geology and Geophysics, Yuzhno-Sakhalinsk-2, 693002, U.S.S.R.

(Received: 5 March 1990, revised: 14 June 1990)

Abstract. Two inexpensive cable bottom pressure stations were installed on the southwestern shelf of Kamchatka (Okhotsk Sea) in 1987 and two more in 1988 to provide longwave measurements in the tsunami frequency band, to investigate the generating mechanism of these waves, and to test the instrumentation. Microfluctuations of atmospheric pressure were recorded simultaneously. Two cable lines were torn off by ship anchors in March 1989 but others are still working in spite of highly dynamic activity on beaches and in hard ice regimes. Careful data analysis of two months of observations (September–October, 1987) showed that: (1) the atmospheric spectra were very stable and monotonic in the period range 2–50 min and corresponded to a power law of $\omega^{-2.3}$, (2) the direct generation of long waves by atmospheric pressure fluctuations was negligible, (3) there was high correlation between background longwave oscillations and sea state, (4) the structure of the offshore longwave field was in good agreement with theoretical estimates of standing waves for a linear slope.

Key words. Tsunami measurements, bottom cable stations, infragravity waves, atmospheric fluctuations.

1. Introduction

The west coasts of Kamchatka are regions which are strongly exposed to the effects of marine natural disasters. Although the Kuril Islands protect the Okhotsk Sea from the direct threat of tsunami waves, some strong tsunamis (e.g. the Kamchatka Tsunami on 5 November 1952 and the Chile Tsunami on 23 May 1960) have penetrated into the region and caused substantial damage. But the most dangerous and ruinous hazards for these coasts are storm surges and storm-generated wind waves. Six to seven strong surges are observed here each year, the most destructive in recent years occurred in November 1980, January 1982, November 1986, and October 1988. During surges, long surf waves run over sandy tongues of land, upon which the majority of villages and industrial activities are located, and destroy many structures (Figure 1). For research to provide trustworthy forecasts of marine hazards for this region, it is necessary to continuously monitor sea-level variations and longwave processes. Unfortunately, the high dynamic activity of beaches makes it difficult to maintain stationary tide gauges on the west coast of Kamchatka and, thus, no systematic sea-level observations are made here. For this reason, the Kamchatka Administration of the Hydrometeorological Service appealed to the

Fig. 1. An example of the destructive effects of sea waves on a nearshore building on the southwestern coast of Kamchatka.

Institute of Marine Geology and Geophysics (IMGG) to make measurements based on bottom cable stations. This capability has been developed at IMGG for the remote sensing of tsunamis and other long waves in a broad frequency band.

The Kamchatka shelf of the Okhotsk Sea is a complex and interesting area that has been poorly investigated. The relatively long and homogeneous shelf, and a remarkably straight and plane coastline are conducive to the formation of boundary long waves associated with both atmospheric activity and the nonlinear interactions of wind waves.

During the IMGG expeditions in 1987–1989, several experiments were conducted on the southwestern shelf of Kamchatka and routine sea level recording was established. The main scientific purposes of these experiments were

(a) instrument testing, and a study fo the feasibility and efficiency of establishing inexpensive, long-term bottom cable stations in regions characterized by high beach instability and severe ice conditions;
(b) the investigation of long-wave background spectra, with special attention to generating mechanisms and correlations with possible external sources;
(c) studies of tide and storm surges in this area and the development of adequate forecast models.

Item (c) is beyond the scope of this paper. Our attention will focus on long waves in the tsunami frequency band, i.e. on waves with periods of 2–120 min. The nature of these waves on the shelf was of great interest in the 1950s and 1960s (Munk *et al.*, 1956, Donn *et al.*, 1956; Takahasi and Aida, 1962; Snodgrass *et al.*, 1962; Munk, 1962; Munk *et al.*, 1964). It was shown that background spectra are dominated by two types of boundary waves: edge waves, in which energy is trapped on the shelf and decays asymptotically seaward, and leaky waves which radiate energy into the open sea. In the absence of tsunamis, the primary forcing mechanism for these waves was assumed to be atmospheric fluctuations. Since the nature and pecularities of background oscillations in the tsunami frequency band have been considered to be fairly well-established, only a few studies related to this frequency band have been published recently (Bernard and Milburn, 1985; Thompson and Van Dorn, 1985; Hashimoto *et al.*, 1986; Van Dorn, 1987).

On the other hand, a great many papers (Huntley *et al.*, 1981; Holman, 1981; Middleton *et al.*, 1987; and others) have been devoted to 'infragravity waves' with periods 30–300 sec, associated with the nonlinear interaction of wind waves and swell in coastal waters. The importance of these waves is due to their role in beach transformation and destruction (Bowen and Huntley, 1984; Holman and Sallenger, 1986).

This report presents the principal results of our recent investigation on the shelf of Kamchatka, during which we have found that these two frequency bands are correlated. In particular, we show that wind waves and swell are an important source of background wave energy in the tsunami frequency band.

2. Instruments

The development of instruments for tsunami wave recording at the shelf zone was begun at IMGG (formerly the Sakhalin Complex Scientific Research Institute) on the initiative of S. L. Soloviev toward the end of the 1960s (Jacque and Soloviev, 1971; Jacque *et al.*, 1972). Bottom cable stations with Vibrotron sensors and coastal analog recorders were used initially. Improved sensors, recorders and cable lines made it possible to obtain one of the first records of tsunami waves in the open ocean on 23 February 1980 (Dykhan *et al.*, 1983).

Third-generation instruments, designed at IMGG, were used on the Kamchatka shelf. These instruments allowed the collection of real-time oceanographic data over a broad frequency band, including tides, surges, tsunamis, seiches, and surf beats. Two key features of the marine data system were the Digiquartz pressure sensors used in the bottom stations and the special electronic functional modules in the shore data recorders. The recorders had 4–8 channels which could be used in any combination to provide an optimal observational system; the system at Ozernovsk, e.g. consisted of three bottom pressure gages and a coastal microbarograph. Some additional details on the instrumentation may be found in Kovalev *et al.* (1989). A full description of the bottom pressure stations and other instrumentation will also be published by IMGG in 1991 as part of a special collection of technical papers.

Two types of bottom pressure sensors were used on the southwestern shelf of Kamchatka. The first was a standard Vibrotron sensor PDV-10, a similar but improved version of that used for tsunami observations near Shikotan Island (Dykhan *et al.*, 1983). The second was a quartz crystal transducer with a range of 0–3 MPa, hysteresis of 0.05%, and a typical accuracy under harsh environmental conditions of 0.25%. Some quartz crystals were also used as temperature gages. In addition, a precision shore microbarograph, designed at the Moscow Institute of Physical Engineering, provided simultaneous measurements of atmospheric fluctuations. The microbarograph also employed a quartz crystal sensor, with a range of 500–1100 hPa and a typical accuracy of 0.03%. Bottom sensor signals were transmitted via armored one-wire cables, and microbarograph signals via ordinary line, to a shore-based recorder and transformed there by an electronic analog filter-amplifier and amplitude selector. The transformed signal from each sensor (channel) was then integrated over a set time interval (0.25–8 min) and written on magnetic tape (compact-cassette) as sequential 15-bit words. The continuous simultaneous recording of ocean and atmospheric waves was intended to clarify the origin of long ocean waves in the tsunami frequency band. The instrument system also included a frequency/analog converter to provide direct visualization of sea-level oscillations as an aid to decisions involving marine hazard warnings.

3. Field Observations

Long-wave field measurements on the shelf of Kamchatka were carried out in September 1987 near Ozernovsk Village in southwest Kamchatka (Experiment KAMSHEL-87). The first cable station (V1) with a Vibrotron bottom pressure sensor was installed on 10 September at a depth of 9.5 m, about 800 m offshore; the second station (Q2) with a quartz sensor at a depth of 17 m, 1.4 km offshore, was installed the next day (Figure 2). The cables were connected to a multichannel recorder deployed in the winter dispatcher building of the collective fish farm, only 30 m from the waterline. No special effort was made to protect the cables in the surf zone, but the placement of the cables and their route through the surf zone and out of the water was carefully chosen. The precision microbarograph was installed in the same building on 1 September. The sampling interval (integration time) for all parameters was chosen to be 1 min.

The microbarograph worked until 7 October 1987. Water wave measurements continued with minor interruptions until 1 April 1988. Periodic changes of magnetic tape cassettes and routine recorder maintenance were performed by fish-farm workers. The measurements were terminated with the seasonal closing of the dispatcher building.

In September 1988, investigations on the Kamchatka shelf were resumed (Experiment KAMSHEL-88) and water wave recording stations V1 and Q2 were reacti-

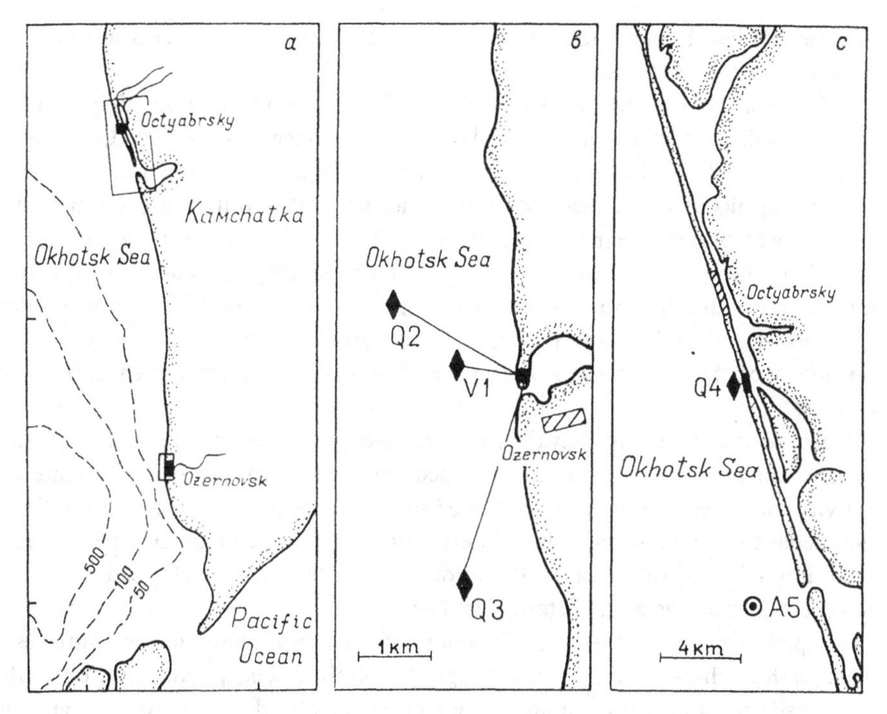

Fig. 2. Location of bottom cable stations on the southwestern shelf of Kamchatka.

vated. Data quality did not deteriorate, despite the one-year submergence of cables and sensors in the sea. In addition, station Q3 with quartz bottom pressure and water temperature sensors was deployed two miles south of station V1 at a depth of 10.5 m (Figure 2b). The corresponding cable was connected to the recorder and data from all stations were written on the same magnetic tape. Unfortunately, one of the farm-workers inadvertently disabled the tape recording at the end of September 1988, and the reason for the failure was not determined until the next IMGG expedition in May 1989. At that time it was also discovered that the cables to stations Q2 and Q3 were ruptured about 1.2 km from the shore. This apparently occurred in March 1989, when ships anchored nearby for shelter from the storm. As an unfortunte consequence, only a few days of simultaneous recordings at all three stations were obtained.

An inspection showed that cable conditions in the surf zone were quite satisfactory. The cables had been covered with sand, which protected them from damage. We also note that there was an early attempt in 1986 to install a cable bottom station with an analog recorder at Ozernovsk to provide tide and storm-surge observations for the Hydrometeorological Service. This attempt failed due to an unfortunate placement of the cable, as the line was destroyed in the surf zone within three weeks. The microbarograph was deployed again in May 1989, and at present

(February 1990) both station V1 and the microbarograph are successfully collecting data with few interruptions.

Additional instruments were also installed near the Octyabrsky Village in September 1988 as part of the KAMSHEL-88 experiment. At station Q4, a quartz sensor was deployed at a depth of 5.5 m about 600 m offshore (Figure 2c). In this region, big blocks of ice heap one upon another in the wintertime. To protect the cable, it was buried in sand to a depth of 2–3 m. As at Ozernovsk, a recorder and microbarograph were located in the dispatcher building. In addition, a self-contained unit with a vibrotron sensor was deployed at station A5 at a bottom depth of 20 m, about 20 km south of Q4. This station worked for three weeks. The sensitivity of the sensor was low and the data was subsequently used only for tide analysis.

Cassette exchanges in Octyabrsky were also performed by the local fishermen. Parallel analog recorder output provided them with tide level and marine-wave activity estimates. Unfortunately, power supply problems and a lack of continuous maintenance caused several data interruptions. At the end of October 1989, the recording of long water waves at station Q4 was temporarily stopped: we plan to resume observations at this station in 1990.

In spite of some misfortune, the results of the experiments on the southwestern shelf of Kamchatka are very encouraging. Relatively cheap cable stations, which were installed in a very complicated region characterized by a hard ice regime and high dynamic beach activity, have been working quite successfully for more than two years. Valuable scientific data have been obtained and are still being recorded. Some results of the data processing are presented below.

4. Data Description and Preliminary Analysis

All information from the cable stations and microbarograph were recorded on magnetic tape, read into a computer, and carefully verified. Each tape stored about 6–7 days of observations. This paper is devoted mainly to the analysis of data collected during the period September–October 1987 (the first 8 tapes obtained in Ozernovsk). In the Kamchatka region, these months are characterized by high activity and strong storms, but coastal waters are not yet ice-bound. Recorder maintenance procedures caused some loss of data between tapes, with the largest interruption of about 25 h occurring between tape numbers 4 and 5. Additional omissions occurred in the records of atmospheric fluctuations because of microbarograph power failure.

Atmospheric pressure and wind data from the Hydrometeorological Station at Ozernovsk were also used in the analysis. The standard time step of these data was 3 h, but station personnel made a special effort to measure these parameters every hour during intense storms, as a much-appreciated favor to the project. The least-square method was used to estimate tidal constituent for the records of stations V1 and Q2, then the predicted tides were subtracted from the initial records. High-pass Kaiser-Bessel filters (Harris, 1978) were then applied to the residual sea-level time series and the atmospheric pressure fluctuation records to isolate the high frequency components. These series were then used in the subsequent analysis.

5. Atmospheric Processes and Their Correlation with Background Longwave Oscillations in the Sea

The Okhotsk Sea is a zone of active cyclonegenesis, and the central part of the Okhotsk is characterized by the maximum number of cyclones occurring in the North Pacific (The Pacific Ocean, 1966). Some storms are very intense here, e.g. a cyclone in March 1988 had a central pressure of 942 mbar and was followed by devastating winds. Most approach Kamchatka from the southwest (i.e., from the Kuril Islands) or the west (from the Okhotsk Sea) (The Pacific Ocean, 1966; Likacheva *et al.*, 1985).

In the original atmospheric pressure records, typical amplitudes of the microfluctuations in calm, high pressure weather were 0.10–0.15 mbar; these amplitudes would increase to 0.3–0.5 mbar during periods of low pressure, with individual wave amplitudes exceeding 1 mbar. We subjected both microbarograph and ocean bottom pressure records to a high-pass Kaiser-Bessel filter with 2.5 h cutoff period. We then computed the rms atmospheric fluctuations, σ_p, and the rms longwave oscillations, σ_ζ, over 3 h segments of the records. These are presented in Figure 3, along with the simultaneous series of 3 hourly atmospheric pressure and wind speed observations.

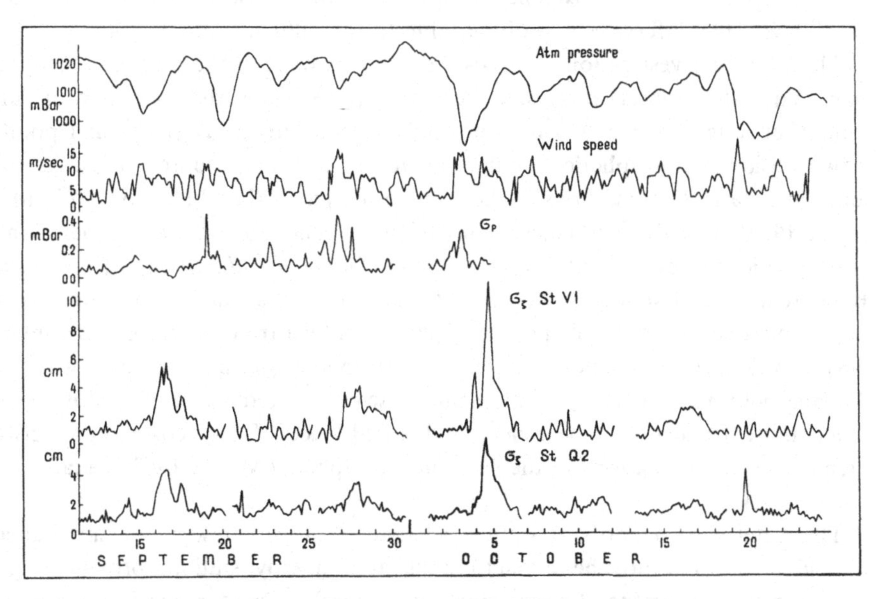

Fig. 3. Time series of atmospheric pressure, wind speed at the Ozernovsk Village, root mean square (rms) variation of atmospheric fluctuations (σ_p) and rms variation of long waves at stations V1 and Q2 (σ_ζ).

During the period under study, there were 5 weak and 4 strong cyclones in the Ozernovsk region. The strongest storms occurred during the periods 14–16 September, 18–21 September (typhoon Freda), 2–6 October, and 19–22 October, i.e. one cyclone every 5–6 days (Figure 3). This cyclone recurrence period is quite typical for the Okhotsk Sea region (Likhacheva and Rabinovich, 1985). Cyclone passage was accompanied by intensification of atmospheric fluctuations, and maximum fluctuations usually occurred at the front of a storm, 12–18 h before the minimum pressure was registered (Figure 3). There is a clear correlation between wind speed and the intensity of high-frequency atmospheric pressure oscillations, strengthening of the wind was usually accompanied by a growth in pressure fluctuations. But there are exceptions. For example, strong winds on 15 and 19 September did not cause noticeable pressure fluctuations; on the other hand, strong atmospheric waves were observed on 22 September, when the wind was relatively weak (Figure 3).

Spectral analyses of all data were then performed, using time series which had first been subjected to a Kaiser-Bessel high-pass filter with a 4 h cutoff period, then tapered with a Kaiser-Bessel window to improve the spectral estimates (Harris, 1978). The results of the atmospheric wave spectral analysis were unexpected. During the observational period (10 September–7 October) the spectra were smoothly monotonic in character and were extremely stable, although total energy varied by 1–1.5 orders, depending on atmospheric activity, the spectral shapes were practically the same in each case, and were best fit by an $\omega^{-2.3}$ power law (Figure 4). This is steeper than observed by other scientists; according to Gossard (1960), Golitsyn (1964), Herron *et al.* (1969), and Kimball and Lemon (1970), atmospheric pressure spectra in the frequency range 10^{-4}–10^{0} Hz decrease according to an $\omega^{-2.0}$ law. This difference is probably due to the influence of the sea.

Although an investigation of atmospheric waves was not the primary objective of this study, their observation was necessary to estimate the efficiency of direct forcing of long ocean waves by atmospheric disturbances. A traditional point of view considers atmospheric fluctuations to be the main source of background longwave sea-level oscillations in the tsunami frequency band (Munk, 1962; Munk *et al.*, 1964), and there is indeed convincing evidence of longwave generation by atmospheric waves (Munk *et al.*, 1956.; Donn *et al.*, 1956). Furthermore, Bondarenko and Bychkov (1983) describe an interesting example of long waves in the Caspian Sea which were apparently generated by a train of atmospheric internal gravity waves; the coherence between the atmospheric and marine waves was about 0.6, and both had periods of 23 min. But these cases are the exception, not the rule; much more frequently, researchers find an absence of any correlation between high-frequency oscillations in the sea and atmosphere (Munk, 1962; Takahasi and Aida, 1962).

The water-wave records at station V1 and Q2, together with the simultaneous measurement of atmospheric fluctuations at a nearby station, provide a good opportunity to examine this question more closely. Cross-spectral analyses were made using record segments of 3–6 days. A fast Fourier transform technique was

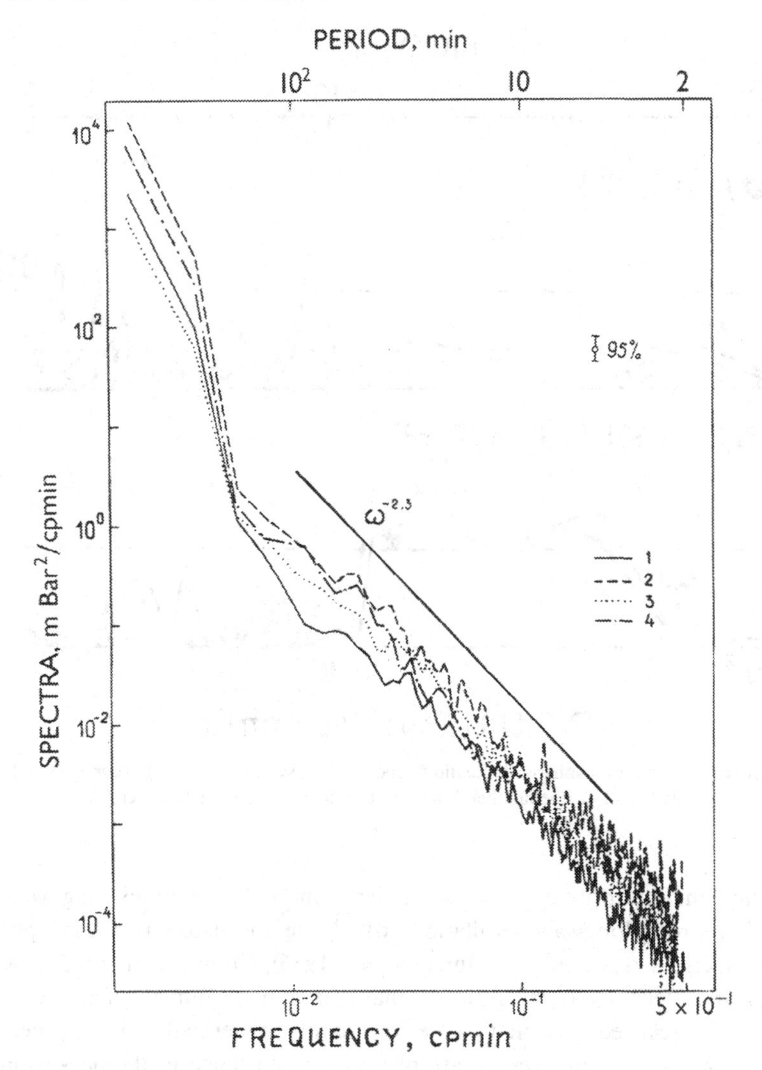

Fig. 4. Spectra of atmospheric pressure fluctuations for 4 different 5-day intervals in September–October, 1987.

used to obtain coherence and phase lag estimates; typically, the number of degrees of freedom was 100–200. For practically all of the period under study, the coherence between sea-level and atmospheric oscillations was less than 0.1, i.e. well below a significant confidence level. Figure 5a presents a very typical example. The single exception was the period 26 September–1 October when the coherence exceeded the confidence level and reached a value of 0.13 in the frequency range 0.02–0.09 cpmin (Figure 5b).

This behavior of the coherence between atmospheric and marine disturbances suggest that direct forcing of the sea surface by atmospheric pressure does not

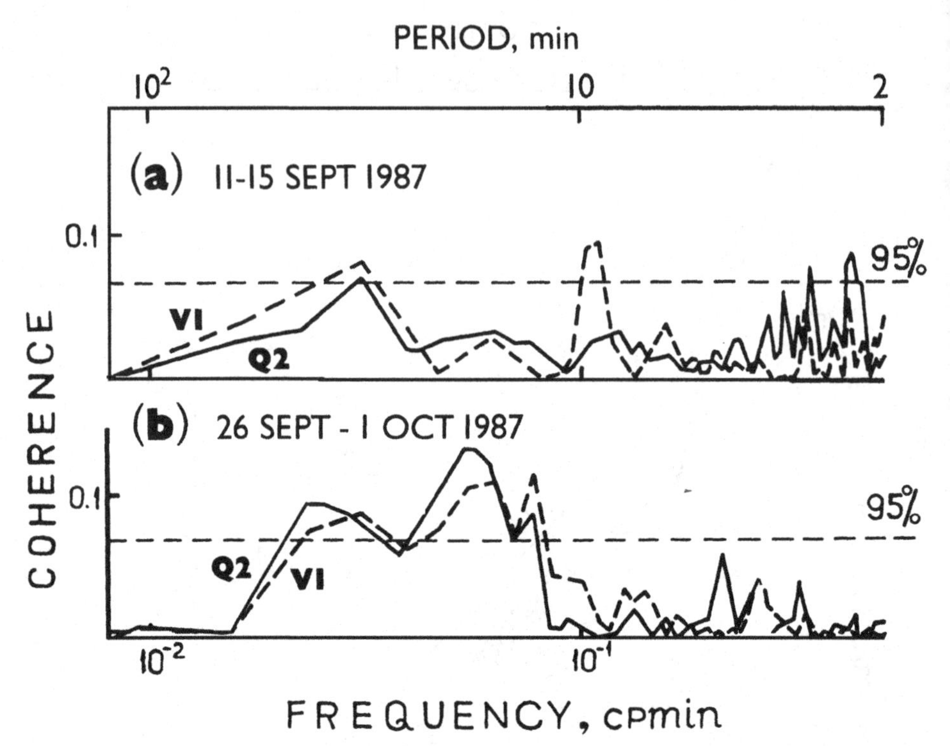

Fig. 5. Coherence between atmospheric fluctuations and long waves at stations Q2 (curve 1) and V1 (curve 2) for the periods (a) 11–15 September 1987 and (b) 26 September–1 October 1987.

normally play an important role in the generation of background long waves. It appears that strong longwave oscillations which are correlated with atmosphere fluctuations, such as observed by Munk *et al.* (1956), Donn *et al.* (1956), and Bondarenko and Bychkov (1983), are resonant in nature. But resonance conditions are rarely realized, so that these resonantly generated waves ('meteo-tsunami'), like actual tsunami waves, are of minor importance in the background spectra.

This being the case, what causes the background longwave oscillations with tsunami periods? There is a definite correlation between time variations of atmospheric pressure and longwave intensity (Figure 3). Typical longwave amplitudes were 1–3 cm during calm weather and from 5–15 cm during cyclone passages. Evidently, the atmosphere is the main source of excitation for these waves, but the energy transfer mechanism is far from trivial. Two mechanisms for indirect longwave generation are possible:

(1) scattering of 'meteotide' (surge sea-level displacement) by bottom and coastline irregularities;
(2) nonlinear interaction of water gravity waves (swell, wind waves).

The first mechanism is well known for internal waves, and irregularities of relief are the main generating source (LeBlond and Mysak, 1977). Kulikov and Shevchenko (1985) showed that surface long waves may also be generated by a similar process. The second mechanism plays an important role in forming surf beats – longwave motions in a coastal zone with time scales of about 30–300 sec (Munk, 1949). Longuet-Higgins and Stewart (1962) described theoretically the origin of a forced wave component of surf beats, connected tightly with the group structure of surface gravity waves; Gallagher (1971) and Bowen and Guza (1978) have demonstrated the possibility of free edge waves generated by wind waves and swell.

Both mechanisms are characterized by a two-cycle energy transfer from the atmosphere to long waves. Which process is more important for waves in the tsunami frequency band is the subject of this special study.

6. The Correlation of Longwave Spectra with Sea State

On 2–6 October 1987 a strong cyclone passed by Ozernovsk. The cyclone was accompanied by significant longwave oscillations; σ_ζ exceeded 7 cm at station Q2 and 11.5 cm at station V1 (Figure 3). This case provided us with an excellent opportunity to investigate the origin of long waves.

A moving cyclone is associated with a sea-level displacement that represents a forced departure from static level appropriate to the 'inverted barometer law'. If such a displacement was the source of long waves through the process of diffraction and scattering, then waves with maximum amplitude should be observed at or before the moment of minimum pressure (maximum surge), since the longwave velocity is much higher than the cyclone speed. But in fact, the maximum waves clearly lagged the cyclone center by several hours (Figure 6). The correlation between pressure and long wave intensity (σ_ζ) indicates an average time lag of about 33 hours.

The reason for this time lag, and the origin of the observed long waves, may be clarified by comparison with the wind and wind-wave variability. Easterly (offshore) winds prevailed at the front of the cyclone (Figure 6), and only small wind-waves (less than 1 m) were observed. After the cyclone center crossed the region of Ozernovsk, the wind became westerly (onshore) and wind-waves then began to grow rapidly. The time of maximum long waves corresponds exactly to the time of maximum wind waves (visually estimated to be higher than 3.5 m).

Spectral analyses of long waves for different sea surface conditions show that longwave energy increased considerably during the storm in the high-frequency band (up to periods of 35–40 min) but was practically unchanged at low frequencies (Figure 7). Similar results have been obtained in the coastal zones of Japan and the Kuril Islands, where strong correlations have been noted between sea-state and background longwave oscillations. Storm-generated waves were found to accompany significant long waves with periods 1–18 min (Takahasi and Aida, 1962),

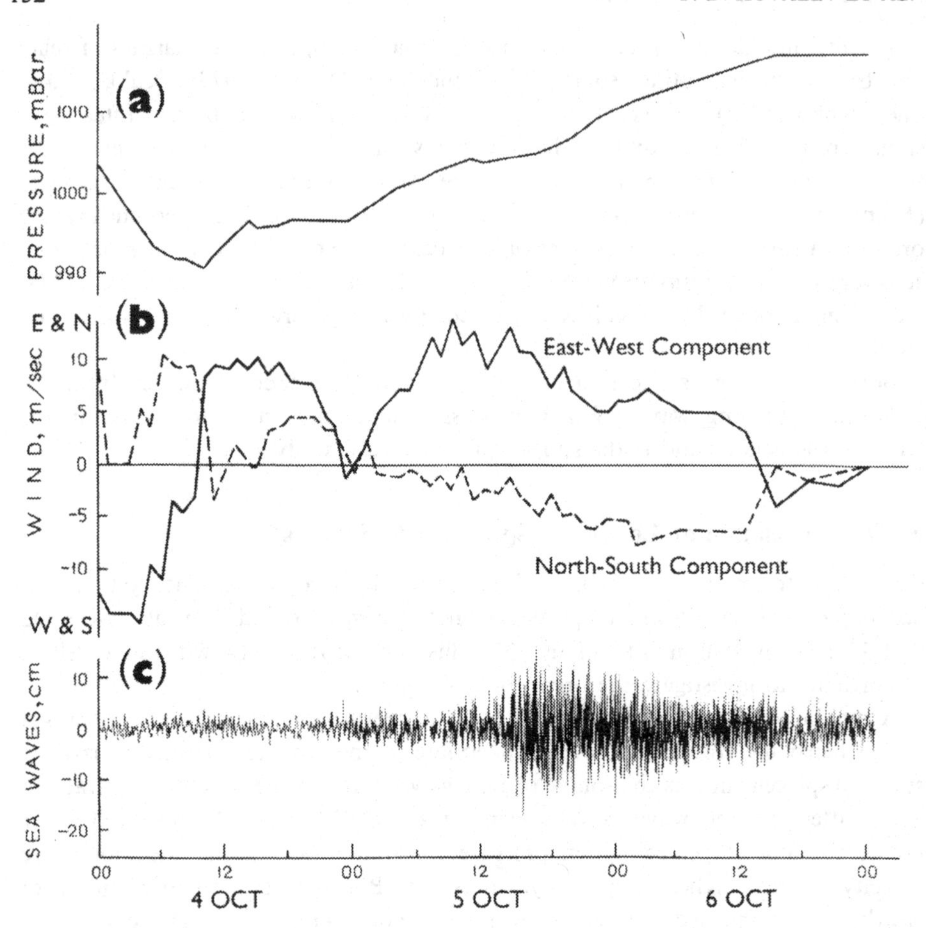

Fig. 6. Variations during cyclone passage on 3–6 October 1987 (a) atmospheric pressure, (b) east-west and north-south wind component at the Ozernovsk Village, and (c) background longwave oscillations at station Q2.

2–25 min (Aida *et al.*, 1970), 1.5–15 min (Bychkov *et al.*, 1972), and 1–25 min (Hashimoto *et al.*, 1986); corresponding spectral bands were 2 orders higher in rough weather than in calm weather.

Such spectral behavior proves that storm passage can cause a phenomenon analogous to 'negative viscosity', a well-known process in turbulence theory (Monin and Ozmidov, 1981), in which motions with high frequencies and small scales transfer energy to large-scale, low frequency movements. Negative viscosity may develop only in the presence of an intensive external source, and energetic storm waves are just such a source. Hasselmann (1971) presented a theoretical mechanism whereby energy could be transferred from short gravity waves to larger-scale motion, and suggested that this may be an important process in the generation of long ocean waves.

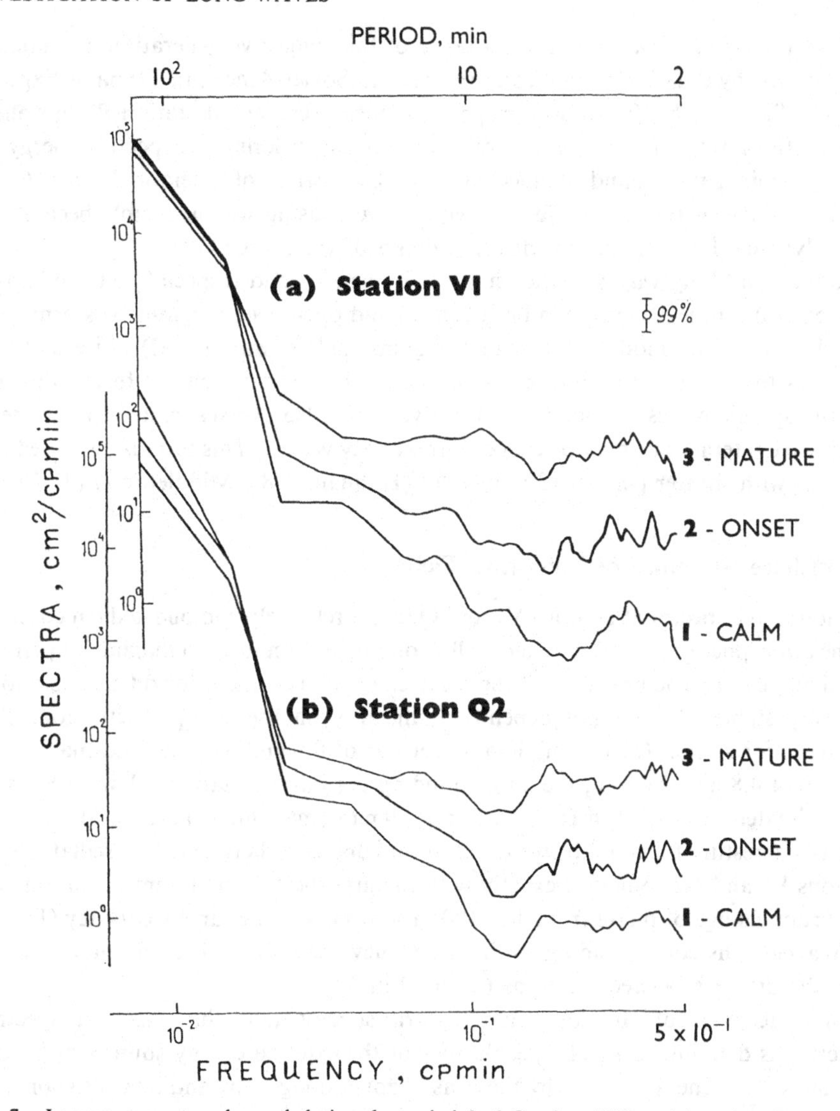

Fig. 7. Longwave spectra observed during the period 3–6 October 1987 at (a) station V1 and (b) station Q2 during three stages of the storm: (1) the calm period before storm arrival, (2) at the onset of the storm, and (3) during the mature phase of the storm.

Unfortunately, the sampling integration period of 1 min precluded quantitative estimates of this process, because of aliasing effects. Specifically, a rectangular filter with a 1 min window does reduce the amplitude of oscillations with a 6–7 sec period by a factor of 30, but that may still be insufficient to remove aliasing effects entirely. Thus, while storm-wave intensification activates nonlinear processes which increase energy transfer to low frequencies, the effects of aliasing also increase simultaneously.

Nevertheless, the reality and importance of this longwave generation mechanism is confirmed by data collected during the Second Soviet-American Tsunami Expedition (Kulikov *et al.*, 1983). Bottom pressure measurements at station P2, installed at a depth of 1000 m, showed a significant increase in longwave spectral energy in the 2–15 min period band, connected with the passage of typhoon Irma (16–18 September 1978). Here, the effect of wind wave aliasing was negligible because of hydrodynamic filtering due to the large depth of the sensor.

Background long waves in the tsunami frequency band (especially at the higher frequencies) can be generated in both coastal and open-ocean regions by storm-generated waves. The model of Longuet-Higgins and Stewart (1962) quite suitably describes the process. In contrast to marine baric waves, which are forced directly by atmospheric waves (Bondarenko and Bychkov, 1983), waves related to nonlinear wind wave interactions may be called 'infragravity waves'. This term is also used for motions with shorter periods (30–300 sec) (Holman, 1981; Middleton *et al.*, 1987).

7. Offshore Structure of Longwave Field

The longwave spectra at stations V1 and Q2 were relatively complicated (in contrast to the atmospheric spectra) and had well-pronounced minima and maxima at periods less than 10 min. The positions of these extremes were different for different stations but very stable and did not depend significantly on the synoptic situation. The spectrum for station Q2 has minima at periods of 8.1 and 3.4 min, and maxima at periods of 4.8 and 2.6 min; the corresponding periods for station V1 were 5.1 and approximately 2.2–2.3 min (minima) and 2.9 min (maximum) (Figure 8).

It is interesting that an abrupt decrease in coherence between the oscillations at stations V1 and Q2 coincided exactly with the first spectral minimum of station Q2; an abrupt change of phase from 0 to 180° also occurs at the same frequency (Figure 8). Wavelengths corresponding to this frequency were about 4–5 km, much longer than the distance between stations (about 1 km).

These features of the longwave spectra suggest that the observed spectral structure is determined not by pecularities of the external energy sources or by the mechanism of generation, but by the coastal zone topography and its corresponding system of eigen-oscillations.

Bathymetric relief in the Ozernovsk region is relatively plane and a linear slope model $h(x) = \alpha x$, where h is the water depth and x is the offshore coordinate, may be used to describe the longwave field in the coastal area. Normally reflected leaky waves are described by the expression (Lamb, 1932):

$$\zeta(\omega; x) = aJ_0(\chi), \tag{1}$$

where

$$\chi = (4\omega^2 x/g\alpha)^{1/2} \tag{2}$$

and where ζ is the sea surface elevation, a is the coastline amplitude, ω is the radial

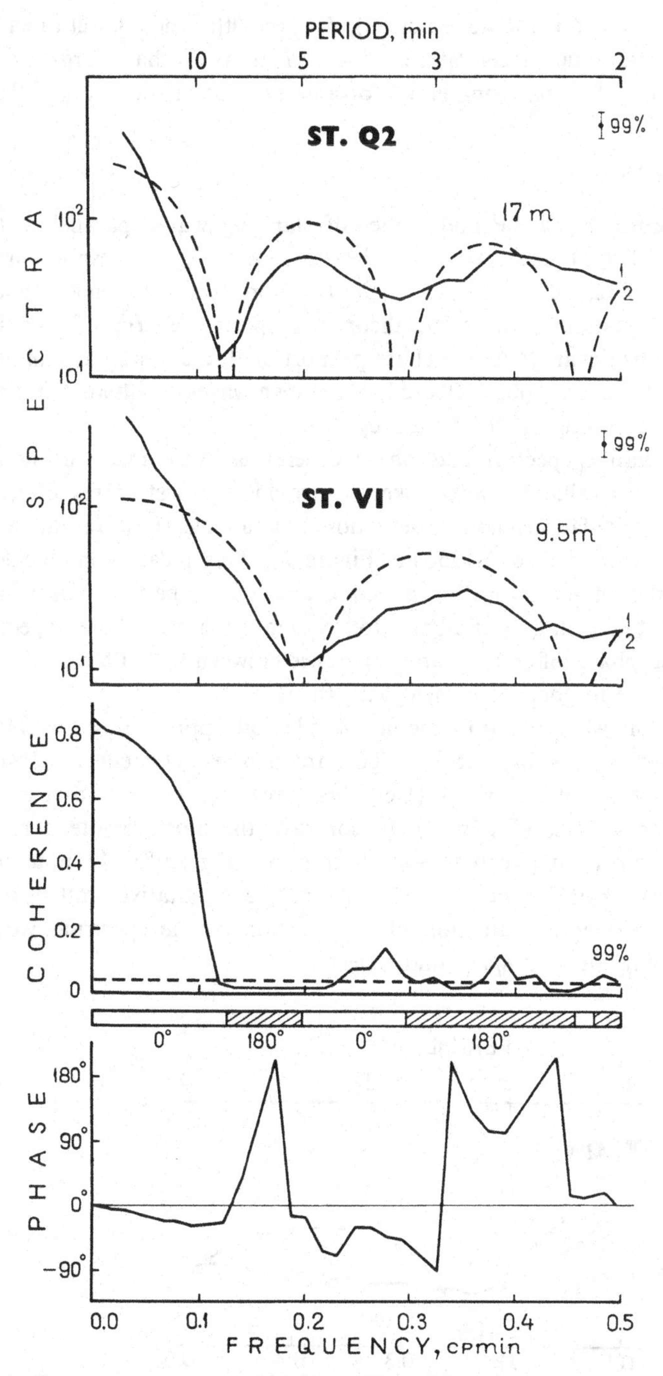

Fig. 8. Longwave spectra at stations Q2 and V1, and the coherence and phase differences between long waves at these stations. Dashed lines are the theoretical spectra for standing long waves; and the theoretical phase is also indicated just below the coherence plot.

frequency ($=2\pi/T$, where T is the wave period), J_0 is the 0th order Bessel function and g is the gravitational acceleration. The function J_0 has zeros when $\chi = \chi_k = 2.405, 5.520, 8.654$, etc. Apparently, for any distance offshore $x = x_i$ there are frequencies

$$\omega_{ki} = \chi_k (g\alpha/4x_i)^{1/2} \tag{3}$$

for which $J_0 = 0$, representing the nodal lines of standing waves parallel to the coastline (Suhayda, 1974; Holman, 1981). At the frequencies ω_{mi}, corresponding to the values $\chi = \chi_m = 3.832, 7.016, 10.174 \ldots$, the function J_0 has extrema that are the antinodes of the standing waves.The theoretical spectral energy of standing waves predicted by this linear slope model are proportional to J_0^2 and have minima (maxima) at the frequencies $\omega_{ki}(\omega_{mi})$. Phase lags between waves at different stations will be 0 or 180°, depending on the frequency.

Calculations of relative spectral and phase differences were made using the expressions (1) and (2); values x_i were picked to provide the best correspondence between theoretical calculations and observational data. The best results were achieved with $x_i = 630$ m and $x_2 = 1250$ m (Figure 8). This means that incident waves were reflected primarily from the breaking line rather than the coast, and that this line was at a distance of about 150–170 m from the shore. Spectral maxima and minima, phase differences, and coherence between the data recorded at stations V1 and Q2 are in good agreement with theory.

The structure of longwave spectra becomes simpler on approaching the shore. The first spectral minimum is displaced to higher frequencies. This effect is clearly seen in the wave spectrum of station Q4 (Octyabrsky region), located closer to the shore than the other stations (Figure 9). In contrast, the most remote offshore station, Q2, has the most complicated spectral structure (Figure 8). We note here that our spectral estimates for periods of 2–3 min are qualitative, rather than quantitative, and a proper investigation of this portion of the spectrum would require denser sampling in both space and time.

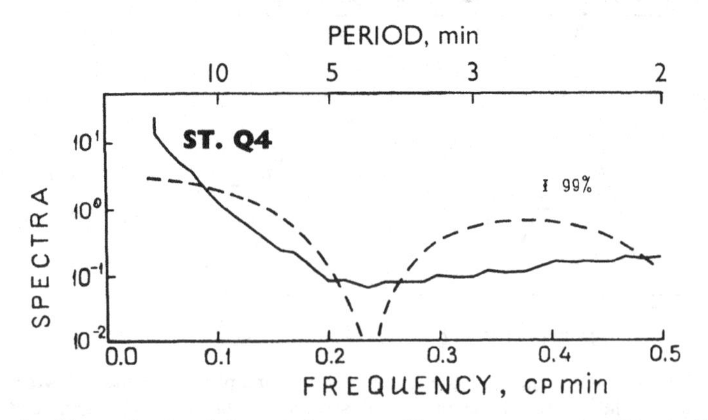

Fig. 9. Observational and theoretical spectra of long waves at station Q4 near Octyabrsky Village.

Similar offshore structure was observed by Suhayda (1974), who investigated wave motions with periods of 1–30 sec. And although he dealt with very different temporal and spatial scales, he too found good agreement with linear shallow water theory for standing leaky waves on a plane beach. These results do not exclude the possibility that edge waves are also present as a component of the wave field. But the profile of a standing edge wave normal to the coast is very similar to that of a normally incident leaky wave of the same frequency, and the separation of these two types of motions would also require a denser instrumental array (Guza and Inman, 1975).

8. Discussion and Conclusions

Measurements of long waves by bottom cable stations on the southwestern shelf of Kamchatka have demonstrated the feasibility of establishing inexpensive, long-term tsunami observation stations in a region with complicated ice, beach and hydrodynamical regimes. Perhaps the most surprising result is that, in spite of widespread opinion that atmospheric waves are the main source of background sea-level oscillations in the tsunami frequency band, these oscillations were found to be weakly correlated with atmospheric fluctuations, but strongly correlated with sea-surface activity. The process of longwave generation is similar to the effect of 'negative viscosity', i.e. energy transfer from small- to large-scale motions, but a thorough investigation of this phenomenon will require continuous measurements of sea waves with a sampling interval of 1–2 sec.

Our results do not eliminate as a source mechanism the resonant generation of long waves by atmospheric disturbance. A search for these apparently infrequent events in existing data sets, and their subsequent analysis, is our next task. Long-term, high quality, simultaneous time series of atmospheric fluctuations and long waves at station V1 were obtained in 1989; these will provide an opportunity for such a study. It will also be interesting to examine the stability of atmospheric spectra and the power law $\omega^{-2.3}$ as a function of different seasons.

A theoretical linear bottom slope model for standing leaky waves was found to be in good agreement with observational data at stations V1 and Q2. Initially, this result may appear to be in contradiction with the results of Munk *et al.* (1964) and Huntley *et al.* (1981), who showed that the background longwave field is composed primarily of trapped edge waves. In fact, however, offshore profiles for standing leaky waves and higher edge wave modes are very similar (Holman, 1981). To separate edge waves and leaky wave modes, and to estimate their relative contribution, it will be necessary to have additional offshore and longshore stations; i.e., a much larger orthogonal array.

Acknowledgements

Vladimir V. Kovbasyuk, from the Kamchatka Hydrometeorological Service, initiated and actively supported this work. Michael P. Puzyrev, Victor V.

Borishchenko, Alexander A. Perederov, and many other fishermen and sailors of Ozernovsk and Octyabrsky assisted in all stages of the experiments, changed tapes, and performed instrument maintenance with genuine interest and sincere goodwill. Frank I. Gonzalez of the U.S.'s National Oceanic and Atmospheric Administration suggested a number of improvements and contributed to the final editing of the manuscript.

References

Aida, I., Date, D., and Koyama, M.: 1970, On the characteristics of long-period fluctuations of the water level observed at Miyagi-Enoshima, *Bull. Earth Res. Inst.* **48**, 983–997 (in Japanese).

Bernard, E. N. and Milburn, H. B.: 1985, Long wave observations near the Galápagos Islands, *J. Geophys. Res.* **90**, 3361–3366.

Bychkov, V.S., Duvanin, A. I., Leybo, A. B., and Monakhov, A. V.: 1972, Temporal evolution of marine baric waves, *Vestnick Moscow Univ., Ser. Geography* **1**, 99-101 (in Russian).

Bondarenko, A. L. and Bychkov, V. S.: 1983, Marine baric waves, *Soviet Meteorol. Hydrol.* **6**, 66–69.

Bowen, A. J. and Guza, R. T.: 1978, Edge waves and surf beat, *J. Geophys. Res.* **83**, 1913–1920.

Bowen A. J. and Huntley, D. A.: 1984, Waves, long waves and nearshore morphology, *Marine Geol.* **60**, 1–13.

Donn, W. L. and Ewing, M: 1956, Stokes' edge waves in Lake Michigan, *Science* **124**, 1238-1242.

Dykhan, B. D., Jacque, V. M., Kulikov, E. A., *et al.*: 1983, Registration of tsunamis in the open ocean, *Marine Geod.* **6**, 303–310.

Gallagher, B.: 1971, Generation of surf beat by non-linear wave interactions. *J. Fluid Mech.* **49,** 1–20.

Golitsyn, G. G.: 1964, On temporal spectra of atmospheric pressure pulsations, *Izv. Akad. Nauk USSR, Ser. Geophysics* **8**, 1253–1258 (in Russian).

Gossard, E.E.: 1960, Spectra of atmospheric scalars, *J. Geophys. Res.* **65**, 3339–3351.

Guza, R. T. and Inman D. C.: 1975, Edge waves and beach cusps, *J. Geophys. Res.* **80**, 2997–3012.

Harris, F. J.: 1978, On the use of windows for harmonic analysis with the discrete Fourier transform, *Proc. IEEE* **66**, 51–83.

Hashimoto, A., Aida, I., Sakashita,S., and Koyama, M.: 1986, On the characteristics of long period fluctuations of the water observed around Oshima Island, *Bull. Earth Res. Inst.* **61**, 129–142 (in Japanese).

Hasselmann, K.: 1971, On the mass and momentum transfer between short gravity waves and large-scale motions, *J. Fluid Mech.* **50**, 189–205.

Herron, T. J., Tolstoy, I., and Kraft, D. W.: 1969, Atmospheric pressure background fluctuations in the mesoscale range, *J. Geophys. Res.* **74**, 1321–1329.

Holman, R. A.: 1981, Infragravity energy in the surf zone, *J. Geophys. Res.* **86**, 6422–6450.

Holman, R. A. and Sallenger, A. H.: 1986, High-energy nearshore processes, *Eos Trans. Amer. Geophys. Union* **67**, 1369–1371.

Huntley, D. A., Guza, R. T., and Thornton, E. B.: 1981, Field observations of surf beat, 1, Progressive edge waves, *J. Geophys. Res.* **86**, 6451–6466.

Jacque, V. M. and Soloviev, S. L: 1971, Distant registration of small waves of tsunami type on the shelf of the Kuril Islands, *Dok. Akad. Nauk USSR* **198**, 816–817 (in Russian).

Jacque, V. M., Velikanov, A. M., and Sapozhnikov, I. N.: 1972, Remote sensing sea-level recorder, *Tsunami Waves, Proc. SakhCSRI* **29**, 189–195 (in Russian).

Kimball, B. A. and Lemon, E. R.: 1970, Spectra of air pressure fluctuations at the soil surface, *J. Geophys. Res.* **75**, 6771–6777.

Kovalev, P. D., Rabinovich, A. B., and Kovbasyuk, V.V.: 1989, Hydrophysical experiment at the southwestern shelf of Kamchatka (KAMSHEL-87), *Oceanology* **29**, (in press).

Kulikov, E. A., Rabinovich, A. B., Spirin, A. I., Poole, S. L., and Soloviev, S. L.: 1983, Measurement of tsunamis in the open ocean, *Marine Geod.* **6**, 311–329.

Kulikov, E. A. and Shevchenko, G. V.: 1985, Generation of edge waves by the meteotide moving along the random borderland, *Theoretical and Experimental Investigations of Long Wave Processes*, FESC, USSR Academy of Sciences, Vladivostok, pp. 20–27 (in Russian).

Lamb, H.: 1932, *Hydrodynamics*, 6th edn., Dover, New York.

LeBlond, P. H. and Mysak, L. A.: 1977, *Ocean Waves*, Elsevier, Amsterdam.

Likhacheva, O. N. and Rabinovich, A. B.: 1986, Estimation of spatial and temporal scales of atmospheric processes in the energy active zones of the World Ocean, *Integrated Global Ocean Monitoring, Proc. 1st Internat. Sympos.*, Tallinn, 2–10 October, 1983, Gidrometeoizdat, Leningrad, Vol. 3, pp. 319–327.

Likhacheva, O. N., Rabinovich, A. B., and Fine, A. V.: 1985, Analysis of atmospheric pressure field over the Okhotsk Sea and northwest Pacific, *Theoretical and Experimental Investigations of Long Wave Processes*, FESC, USSR Academy of Sciences, Vladivostok, pp. 144–157 (in Russian).

Longuet-Higgins, M. S. and Stewart, R. W.: 1962, Radiation stress and mass transport in gravity waves, with application to 'surf-beats', *J. Fluid Mech.* **13**, 481–504.

Middleton, J. H., Cahill, M. L., and Hsieh, W. W.: 1987, Edge waves on the Sydney coast, *J. Geophys. Res.* **92**, 9487–9493.

Monin, A. S. and Ozmidov, R. V.: 1981, *Oceanic Turbulence*, Gidrometeoizdat, Leningrad (in Russian). English edition: *Turbulence in the Ocean*, D. Reidel, Dordrecht, 1985.

Munk, W. H.: 1949, Surf beats, *Trans. Amer. Geophys. Union* **30**, 849–854.

Munk, W. H., Snodgrass, F. E., and Carrier, G. F.: 1956, Edge waves on the continental shelf, *Science* **123**, 127–132.

Munk, W. H.: 1962, Long ocean waves, in *The Sea: Ideas and Observations in Progress in Study of the Sea*, Interscience, New York, pp. 647–663.

Munk, W. H., Snodgrass, F. E., and Gilbert, F.: 1964, Long waves on the continental shelf: an experiment to separate trapped and leaky modes, *J. Fluid Mech.* **20**, 529–554.

Suhayda, J. N.: 1974, Standing waves on beaches. *J. Geophys. Res.* **79**, 3065–3071.

Takahasi, R. and Aida I,: 1962, Spectral analyses of long period ocean waves observed at Izu-Oshima, *Bull. Earth Res. Inst.* **40**, 561–573 (in Japanese).

The Pacific Ocean: Meteorological Conditions over the Pacific Ocean, 1966: Nauka, Moscow (in Russian).

Thompson, W. B. and Van Dorn, W. G.: 1985, Coastal response to tsunamis. *Proc. Internat. Tsunami Sympos., IUGG*, 6–9 August, Sidney, B. C., Canada, pp. 254–263.

Van Dorn, W. G.: 1987, Tide gage response to tsunamis. Part II: Other oceans and smaller seas. *J. Phys. Oceanogr.* **17**, 1507–1516.

Natural Hazards **4**: 161–170, 1991.

Tsunamis In and Near Greece and Their Relation to the Earthquake Focal Mechanisms

B. C. PAPAZACHOS and P. P. DIMITRIU
Geophysical Laboratory, University of Thessaloniki, P.O. Box, 352-1, GR 54006, Thessaloniki, Greece

(Received: 8 December 1989; revised: 3 May 1990)

Abstract. The major earthquake-induced tsunamis reliable known to have occurred in and near Greece since antiquity are considered in the light of the recently obtained reliable data on the mechanisms and focal depths of the earthquakes occurring here. (The earthquake data concern the major shocks of the period 1962–1986.) First, concise information is given on the most devastating tsunamis. Then the relation between the (estimated) maximum tsunami intensity and the earthquake parameters (mechanism and focal depth) is examined. It is revealed that the most devastating tsunamis took place in areas (such as the western part of the Corinthiakos Gulf, the Maliakos Gulf, and the southern Aegean Sea) where earthquakes are due to shallow normal faulting. Other major tsunamis were nucleated along the convex side of the Hellenic arc, characterized by shallow thrust earthquakes. It is probably somewhere there (most likely south of Crete) that the region's largest known tsunami occurred in AD 365, claiming many lives and causing extensive devastation in the entire eastern Mediterranean. Such big tsunamis seem to have a return period of well over 1000 years and can be generated by large shallow earthquakes associated with thrust faulting beneath the Hellenic trench, where the African plate subduces under the Euroasian plate. Lesser tsunamis are known in the northernmost part of the Aegean Sea and in the Sea of Marmara, where strike-slip faulting is observed. Finally, an attempt is made to combine the tsunami and earthquake data into a map of the region's main tsunamigenic zones (areas of the sea bed believed responsible for past tsunamis and expected to nucleate tsunamis in the future).

Key words. Tsunami, tsunami earthquake, earthquake mechanism, tsunamigenic zone, Greece, eastern Mediterranean.

1. Introduction

About 70 major tsunamis, some of them disastrous, are known to have occurred in and near Greece since 479 BC, the year when a big sea wave (the oldest reliably known tsunami) reportedly destroyed the Persian fleet at Potidaea, northern Greece (e.g., see Antonopoulos, 1973). The most devastating of all known tsunamis in the region, the sea wave of AD 365, caused the loss of thousands of lives and extensive damage in the whole eastern Mediterranean and is probably one of the world's largest. Therefore, tsunamis are a real and major hazard to the lives and well-being of the population living along the coasts of the eastern Mediterranean. Hence, the need to assess and mitigate tsunami hazard in this region. As all but a few of the tsunamis here have been generated by earthquakes (one or two major tsunamis resulted from eruptions of Santorini volcano), neither of the above goals can be fully achieved without understanding the mechanism of tsunami nucleation by earthquakes.

The research so far has shown that earthquake parameters as mechanism, moment, and focal depth play a crucial part in tsunami nucleation, but the detailed mechanism of tsunami generation by earthquakes has yet to be understood. There are two main reasons for this. First, only a few large tsunamis have occurred since the birth of instrumental seismology. And, second, only very recently have seismic data become reliable in many tsunami-prone regions.

For the area under investigation, only seismic data since 1962 (the year when the network of standardized long-period stations began to operate) can be considered reliable. In this study, we use the most reliable earthquake data (fault-plane solutions and focal depths), as obtained by Papazachos and his coworkers (see Papazachos, 1988, for a review, and Karacostas, 1988), to shed new light on historic tsunamis and to better assess tsunami hazard in and near Greece.

2. Tsunamis In and Near Greece

Greece and surrounding areas are known for their high seismic activity. This part of the Mediterranean has also experienced several devastating tsunamis. The first complete catalogues of tsunamis in the eastern Mediterranean were compiled by Galanopoulos (1960) and Ambraseys (1962). Antonopoulos (1980) updated these catalogues and enriched them with information from Greek, Byzantine, Arabic, and Latin texts. Papazachos *et al.* (1986a) published a catalogue containing only the reliably known major tsunamis along with the names and epicentral distances of the sites that experienced an estimated tsunami intensity III or larger on the Sieberg–Ambraseys six-degree scale (see Ambraseys, 1962).

We shall now briefly describe the more important, in our opinion, of the tsunamis that have occurred in and near Greece since antiquity.

The oldest historically documented (mentioned in a reliable ancient source) sea wave in Greece is the one that reportedly destroyed the Persian fleet at Potidaea, western Chalkidiki, in 479 BC. According to Bolt (1978), this is the world's oldest historically documented tsunami.

The biggest tsunami observed in and near Greece was due to a presumably very large ($M_s > 8.0$) earthquake in AD 365. We believe the Hellenic trench south of Crete to be the most likely site of the earthquake and of the devastating tsunami it produced. (We shall later present our reasons for reconsidering the site of this event as given in Figure 6 of Papazachos *et al.*, 1986a.) This tsunami claimed thousands of lives and caused widespread devastation in various parts of the eastern Mediterranean (from Crete to Peloponnesus to Alexandria and to Sicily; see Antonopoulos, 1973).

Also disastrous were the earthquake-induced tsunamis in Maliakos Gulf in 426 BC, in the western part of Corinthiakos Gulf in 373 BC and AD 1402 and the tsunami following a submarine eruption of volcano Santorini in 1650.

The largest of the recent tsunamis is the Amorgos tsunami, due to a $M_s = 7.5$ earthquake that hit the southern Aegean area on 9 July 1956 and had its epicenter between Amorgos and Astypalea islands. This tsunami has been thoroughly investigated (e.g., Galanopoulos, 1957; Papazachos *et al.*, 1985). The sea-wave height reached 25 m at the southern coast of Amorgos island, 20 m at the northern coast of Astypalea island, and 2.6 m at the eastern coast of Crete. The damage was considerable on several islands in the southern Aegean Sea, particularly on Kalimnos, where 3 persons were reported drowned.

3. The Mechanisms of the Tsunami Earthquakes

Today, there is little doubt that earthquake mechanism, along with other factors such as earthquake moment and focal depth, plays a critical part in tsunami generation. Therefore, knowledge of the mechanism of a seismic source is crucial in estimating its tsunamigenic potential.

Figure 1 shows the most reliable of the available mechanisms (fault-plane solutions) of the earthquakes with $M_s \geqslant 6.0$ that occurred in and near Greece from 1962 (the year when the network of standardized long-period stations began to operate) to 1986. The black quadrants in the familiar 'bubble' symbols denote pressure and the white ones tension. Thus, symbols with black quadrants in the center indicate thrust faulting caused by horizontal compression, whereas symbols with white quadrants in the center indicate normal faulting produced by horizontal tension. Cases where all four quadrants meet close to the center of the symbol indicate strike-slip faulting.

Tsunamis were produced by the 9 July 1956 Amorgos ($M_s = 7.5$), 6 July 1965 Corinthiakos Gulf ($M_s = 6.9$), 19 February 1968 Hag. Eustratios ($M_s = 7.1$), 15 April 1979 Montenegro ($M_s = 7.1$), 24 February 1981 Alkionidon Gulf ($M_s = 6.7$), and 6 August 1983 Lemnos ($M_s = 7.0$) events, with mechanisms of normal (the first, second and fifth events), thrust (the fourth event) and strike-slip (the third and sixth events) faulting. The respective maximum tsunami intensities were V, II+, III, IV, II and II+ on the Sieberg–Ambraseys six-degree scale (see Ambraseys, 1962).

Apparent in Figure 1 is the grouping of the earthquakes (tsunami and ordinary) according to their mechanisms, which is linked to – and reveals – the main tectonic features of the region (see Papazachos, 1988). Thus, thrust faulting is observed along the outer (convex) side of the Hellenic arc (in agreement with the view that the eastern Mediterranean lithosphere subduces under the Aegean lithosphere) except in its northernmost part (Cefalonia, Leukada), where strike-slip faulting with thrust component is observed and attributed to a dextral transform fault (Scordilis *et al.*, 1985).

The thrust faulting continues further to the north, along the coasts of central mainland Greece, Albania, and Yugoslavia, but this time without evidence of subduction (no Benioff zone), suggesting collision between two continental lithospheric plates (Eurasian–Apulian).

Normal faulting is observed in the whole inner part of the Aegean area, from Crete in the south to central Bulgaria in the north and from eastern Albania and

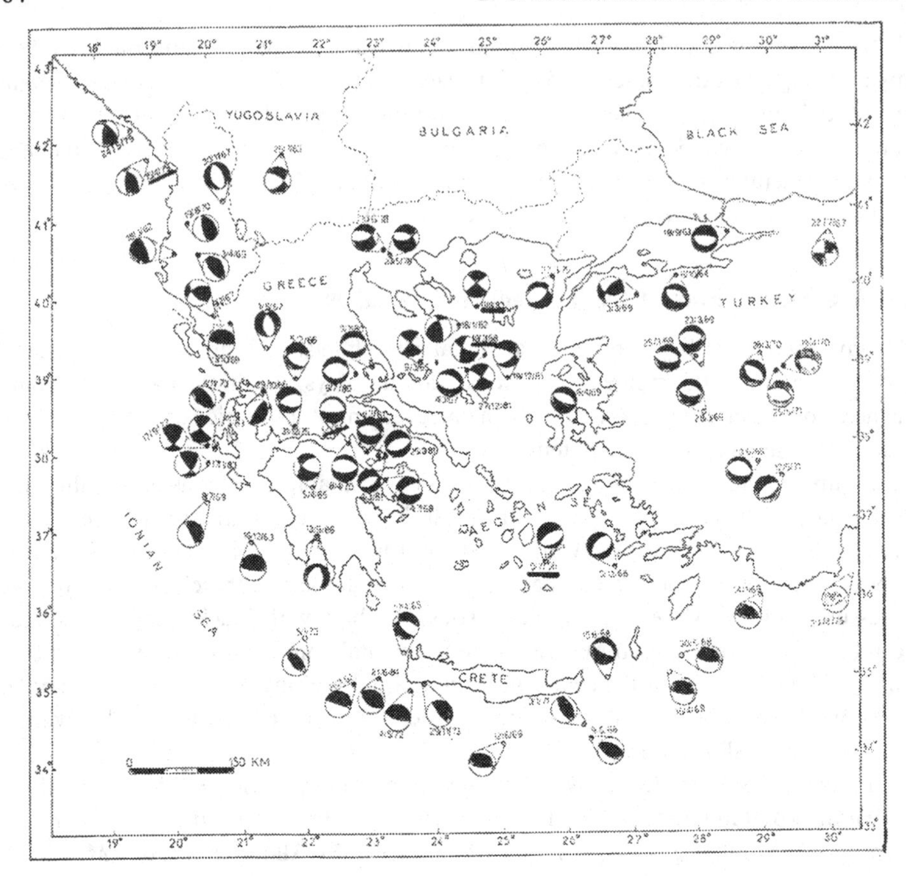

Fig. 1. The most reliable of the available fault-plane solutions for earthquakes with $M_s \geqslant 6.0$ that occurred in and near Greece from 1962 to 1986. The 9/7/1956 Amorgos tsunami earthquake is included with the mechanism proposed by Shirokova (1972). The dates of the tsunami earthquakes are underlined. (Reproduced from Papazachos *et al.*, 1986b.)

central Greece in the west to the whole western Turkey in the east, except in the northwestern part of Turkey and the northernmost part of the Aegean Sea, where strike-slip dextral faulting with either thrust or normal component occurs.

As mentioned earlier, focal depth is another crucial factor in tsunami production. Indeed, observational evidence, supported by theoretical studies, suggests that only shallow submarine earthquakes can produce tsunamis. To help define the main tsunamigenic zones of Greece and surrounding areas and assess their tsunamigenic potential, we present here a map (Figure 2) of the epicenters of the shallow (focal depth up to 60 km) shocks with $M_s \geqslant 6.0$ that occurred in the region between 600 BC and 1986. Figure 2 shows that, apart from the intermediate-depth shocks, defining an amphitheatrically-shaped Benioff zone beneath the southern Aegean Sea, the region's seismicity is characterized by shallow events (the great majority of them with foci within the crust's top 20 km; see Karacostas, 1988).

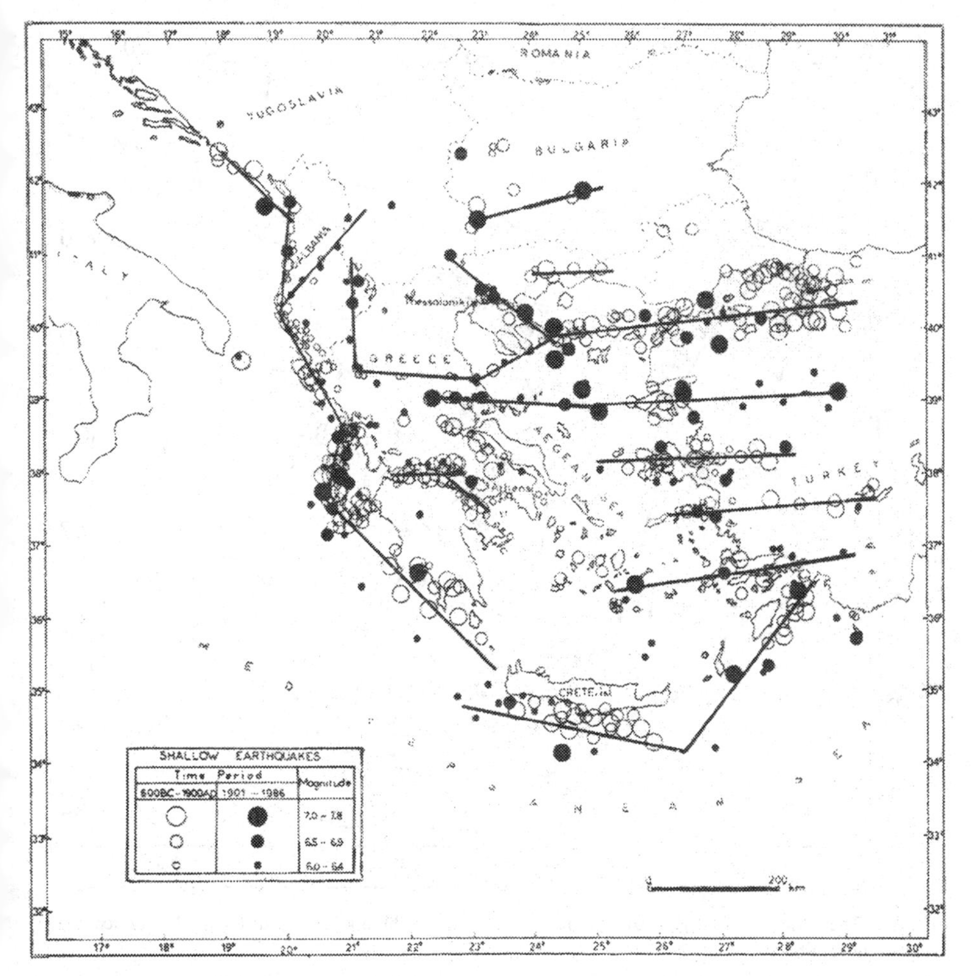

Fig. 2. Epicenters of historic (empty circles) and present-century (filled circles) shallow shocks in and near Greece and the inferred seismic fracture zones. Surface-wave magnitudes are used in the legend. (Reproduced from Papazachos *et al.*, 1986b.)

4. Discussion and Conclusion

The existing tsunami catalogues (e.g., see Papazachos *et al.*, 1986a) imply that only earthquakes with $M_s \geqslant 6.5$ can induce major tsunamis in the broad Aegean area. The most devastating were the tsunamis nucleated in Maliakos Gulf (426 BC), in the western Corinthiakos Gulf (373 BC and AD 1402), south of Crete (AD 365) and at Santorini (1650) and Amorgos (1956) islands. As we shall see, the available tsunami data agree well with the new reliable earthquake data, and both sets of data harmonize with the seismotectonic model of the region proposed by Papazachos and Comninakis (1971) (see Figure 3 and Papazachos, 1988).

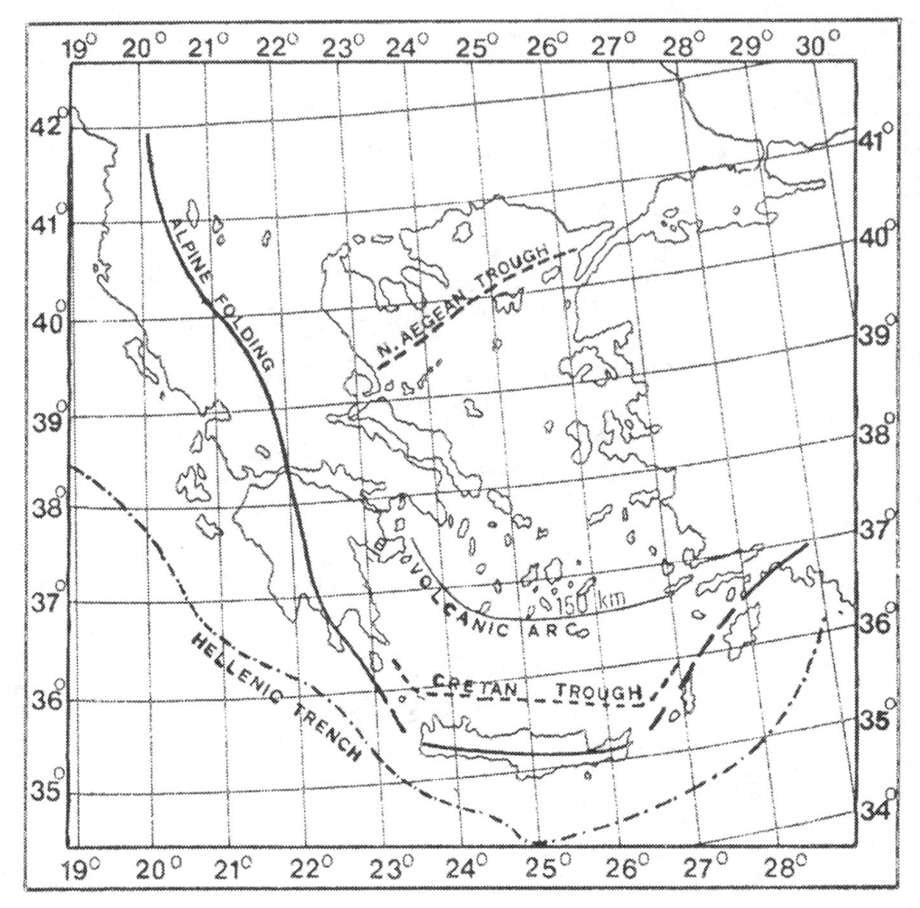

Fig. 3. The main morphologic features of tectonic origin in the broad Aegean area. (Reproduced from Papazachos, 1988.)

According to this model, the Aegean Sea is a typical marginal sea, surrounded by a volcanic arc (Methana–Santorini–Nissiros) and, at a mean distance of some 120 km, by a sedimentary arc consisting of Palaeozoic to Tertiary rocks and constituting a link between the Dinaric Alps and the Turkish Taurides. The volcanic and the sedimentary arcs, collectively known as the Hellenic arc, are paralleled by the Hellenic trench, a series of depressions with water depth to about 5 km. It is along the southern part of the Hellenic trench that the front part of the African plate subduces under the front part of the Eurasian plate. Other prominent morphological features of tectonic origin are the Mediterranean ridge (a submarine crustal swell embracing the Hellenic trench and extending from the Ionian Sea to Cyprus) and the northern Aegean and Cretan troughs, with water depths to about 1.5 and 2 km, respectively.

The above model and the new earthquake data suggest that the largest reliably known tsunami to hit the region, the tsunami of AD 365, had its source somewhere along the axis of the Hellenic trench (presumably south of Crete), rather than north of Crete, as thought earlier (see Papazachos *et al.*, 1986a). Indeed, the seismotectonic conditions in the collision zone between the African and Euroasian plates seem particularly favourable for tsunami generation: large thrust earthquakes with shallow foci occur here. On the other hand, the large earthquakes occurring north of Crete seem incapable, despite having an appropriate mechanism (normal faulting, see Figure 2), of nucleating major tsunamis because of their deep (70 km and deeper) foci. (For example, the $M_s = 8.2$ earthquake of 12 October 1856 produced no sea wave.)

One may then assume thrust faulting as the mechanism of the AD 365 earthquake, which is assigned a magnitude as large as 8.2 because of the severity and geographic extent of the damage it caused (it destroyed at least 10 cities on Crete and possibly many others on the continent and was felt as far as Dalmatia, Sicily, Lebanon, and a part of Egypt and Palestine). As for the tsunami, the scarcity of the available information makes it difficult for one to judge whether the above magnitude is large enough to explain its size. Therefore, one cannot rule out other genetic mechanisms, such as large submarine slumps or the mechanism proposed by Kanamori (1972) to explain the 1896 Sanriku and the 1946 Aleutian tsunamis, initiated by shocks beneath the inner margins of the Japan and Aleutian trenches, respectively. In fact, the similar tectonic conditions of the three regions (Japan, the Aleutian islands, and the eastern Mediterranean) suggest similar genetic mechanisms of the three great tsunamis, but at present we lack evidence (such as the occurrence of large normal-fault shocks along the outer margin of the Hellenic trench) to support this hypothesis.

Among the remaining large tsunamis in the broad Aegean area, all but one (the tsunami due to an eruption of Santorini volcano in 1650) were produced in the back-arc Aegean area (two in Corinthiakos Gulf, one in Maliakos Gulf and one near Amorgos island). This is not surprising, as shallow normal-fault seismic activity is typical of this region (see Figures 1 and 2).

On the other hand, it was recently revealed that the seismic activity in the Ionian Sea (along Cefalonia and Leukada) and in the northern Aegean Sea is mainly due to strike-slip faulting (see Papazachos, 1988 and Figure 1), explaining why no major tsunamis are known to have occurred there despite the occurrence of large earthquakes.

In conclusion, we found that the tsunami data generally agree with the recently obtained reliable earthquake data (focal mechanisms and depths), and both sets of data harmonize with the region's seismotectonic model proposed by Papazachos and Comninakis (1971). We can now proceed with the definition of the main tsunamigenic zones in and around Greece – a first step in the assessment of the tsunami hazard in this region (for details, see Papazachos *et al.*, 1986a).

We call an area of the sea bed presumably responsible for a past tsunami (or tsunamis) and likely to nucleate tsunamis in the future, a *tsunamigenic zone* (or source); its shape is assumed elliptic, the size is proportional to the (estimated) maximum tsunami intensity, and the orientation is taken along the corresponding seismic fracture zone (see Figure 2). Because of the lack of information on past tsunamis and their genetic earthquakes, the relation between the size of the tsunami source and the maximum tsunami intensity is found on the basis of the observation that the sea-bed area responsible for a tsunami is approximately equal to the aftershock area of the genetic earthquake (see Iida, 1958; Hatori, 1969 and 1981). Thus, the length of the major axis of a tsunamigenic zone is evaluated from the

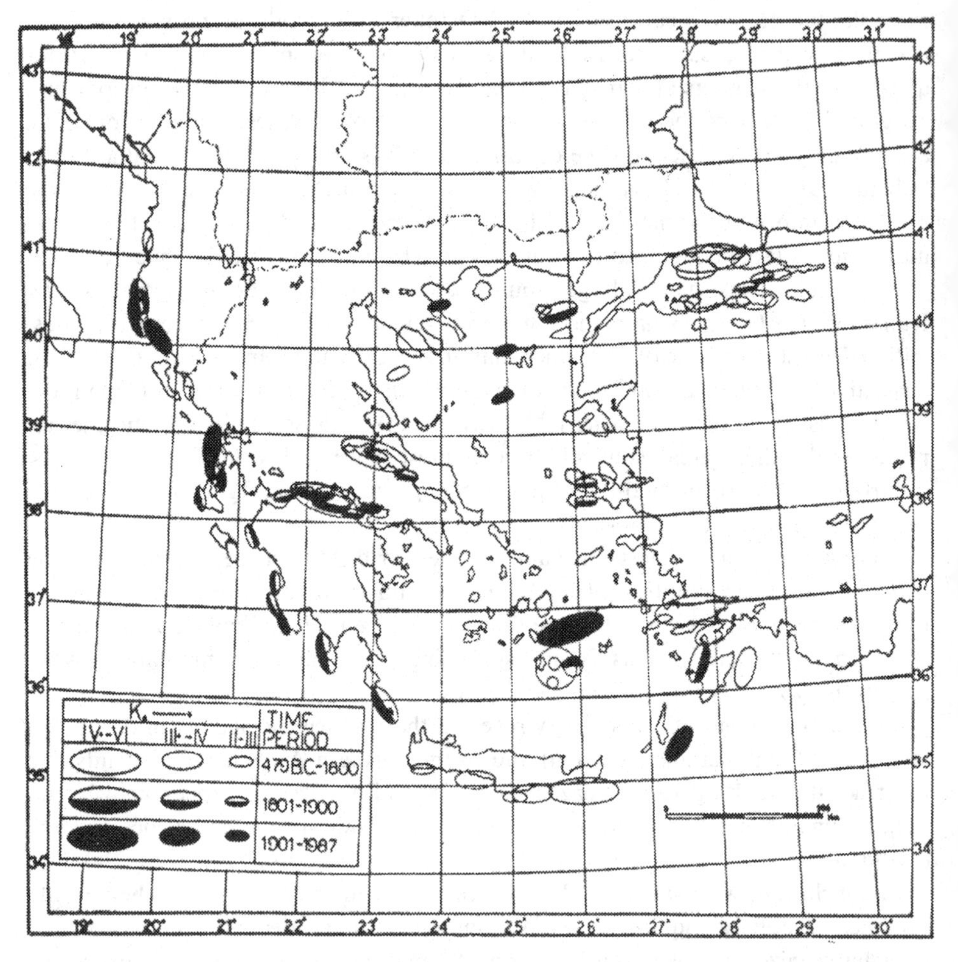

Fig. 4. Known and inferred tsunamigenic zones (sources) in and near Greece. The zones represent areas of the sea bed believed responsible for past tsunamis and expected to nucleate tsunamis in the future. Their shape is assumed elliptic, with the size proportional to the (estimated) maximum tsunami intensity, and the orientation of each zone is taken along the corresponding seismic fracture zone (inferred from the distribution of shallow-shock epicenters; see Figure 2). K_0 in the legend is the maximum tsunami intensity on the 6-degree Sieberg–Ambraseys scale (see Ambraseys, 1962). The circles represent tsunamigenic eruptions of volcano Santorini.

correlational relation between the length, L, of the aftershock area and the magnitude, M_s, of the mainshock for seismic sequences in and near Greece (Karakaisis, 1984),

$$\log L = -2.22 + 0.57\, M_s,$$

after replacing M_s by the maximum tsunami intensity, K_0, from the correlational relation linking the two quantities (Papazachos *et al.*, 1986a)

$$K_0 = (15.92 \pm 4.08) + (2.83 \pm 0.73)\, M_s.$$

The equation relating the length of the major axis of the tsunamigenic zone to the (estimated) maximum tsunami intensity has then the form:

$$\log L = 0.99 + 0.20\, K_0.$$

(The minor axis of the tsunamigenic zones is taken equal to $L/3$.)

The map of the tsunamigenic zones (sources) of Greece and surrounding areas, Figure 4, summarizes the results of our study. Thus, as particularly tsunami-prone should be regarded the regions of the Corinthiakos and Maliakos Gulfs and the southern Aegean area. But the greatest threat comes, we believe, from the tsunamigenic zone that produced the great tsunami of AD 365, which devastated the coasts of the eastern Mediterranean, taking thousands of lives. We locate the source of this tsunami south of Crete, beneath the Hellenic trench, where the African plate subduces under the Euroasian plate, and estimate its repeat time to be of the order of 1000 years.

Acknowledgements

This work is our first contribution to the project 'Assessment and Mitigation of Tsunami Hazard in the Afro-European Tsunamigenic Belt', financed by World Laboratory, and was presented at the International Tsunami Symposium held in Novosibirsk from 31 July to 3 August 1989. We thank Stefano Tinti, the head of the project, and World Laboratory for the financial support that enabled the second author to attend the symposium in Novosibirsk. We also thank the anonymous reviewers for their comments and suggestions.

References

Ambraseys, N. N.: 1962, Data for the investigation of the seismic sea waves in the eastern Mediterranean, *Bull. Seismol. Soc. Am.* **52**, 895–913.

Antonopoulos, J. A.: 1973, *Tsunamis of Eastern Mediterranean from Antiquity until Today*, Athens (in Greek).

Antonopoulos, J. A.: 1980, Data from the investigation of seismic sea-wave events in the eastern Mediterranean from the birth of Christ to 1980 AD (six parts), *Ann. Geophys.* **33**, 141–248.

Bolt, B. A.: 1978, *Earthquakes*, W. H. Freeman, San Francisco.

Galanopoulos, A. G.: 1957, The seismic sea wave of July 9, 1956, *Prakt. Acad. Athens* **32**, 90–101 (in Greek).

Hatori, T.: 1969, Dimensions and geographic distribution of tsunami sources near Japan, *Bull. Earthq. Res. Inst.* **47**, 185–214.

Hatori, T.: 1981, Tsunami magnitude and source area of the Aleutian-Alaska tsunamis, *Bull. Earthq. Res. Inst.* **56**, 97–110.

Iida, K.: 1958, Magnitude and energy of earthquakes accompanied by tsunamis and tsunami energy, *J. Earth Sciences, Nagoya Univ.* **4**, 1–43.

Kanamori, H.: 1972, Mechanism of tsunami earthquakes, *Phys. Earth Plan*et. *Interiors* **6**, 346–359.

Karacostas, B. G.: 1988, Relationship of the seismic activity with geologic and geomorphologic data of the wide Aegean area, PhD thesis, Univ. of Thessaloniki (in Greek).

Karakaisis, G. F.: 1984, Contribution to the study of the seismic sequences in the Aegean and surrounding areas, PhD thesis, Univ. of Thessaloniki (in Greek).

Papazachos, B. C. and Comninakis, P. E.: 1971, Geophysical and tectonic features of the Aegean arc, *J. Geophys. Res.* **76**, 8517–8533.

Papazachos, B. C., Koutitas, Ch., Hatzidimitriou, P. M., Karacostas, B. G., and Papaioannou, Ch. A.: 1985, Source and short-distance propagation of the July 9, 1956 southern Aegean tsunami, *Marine Geol.* **65**, 343–351.

Papazachos, B. C., Koutitas, Ch., Hatzidimitriou, P. M., Karacostas, B. G., and Papaioannou, Ch. A.: 1986a, Tsunami hazard in Greece and the surrounding area, *Ann. Geophys.* **4**, 79-90.

Papazachos, B. C., Kiratzi, A. A., Hatzidimitriou, P. M., and Karacostas, B. G.: 1986b, Seismotectonic properties of the Aegean area that restrict valid geodynamic models, 2nd Wegener Conference, Dionysos, Greece, 14–16 May 1986, pp. 1–16.

Papazachos, B. C.: 1988, Active tectonics in the Aegean and surrounding area, in J. Bonnin *et al.* (eds), *Seismic Hazard in Mediterranean Regions*, Kluwer Academic Publishers, Dordrecht, pp. 301–331.

Scordilis, E. M., Karakaisis, G. F., Karacostas, B. G., Panagiotopoulos, D. G., and Papazachos, B. C.: 1985, Evidence for transform faulting in the Ionian Sea: The Cefalonia island earthquake sequence of 1983, *Pageoph* **123**, 387–397.

Shirokova, E. I.: 1972, Stress pattern and probable motion in the earthquake foci of the Asia-Mediterranean seismic belt, in L. M. Balakina *et al.* (eds), *Elastic Strain Field of the Earth and Mechanisms of Earthquake Sources*, Nauka, Moscow.

Natural Hazards **4**: 171–191, 1991.

Numerical Simulation of Tsunamis – Its Present and Near Future

N. SHUTO
Department of Civil Engineering, Tohoku University, Aoba, Sendai 980, Japan

(Received: 15 March 1990; revised: 9 July 1990)

Abstract. Hindcasting of a tsunami by numerical simulations is a process of lengthy and complicated deductions, knowing only the final results such as run-up heights and tide records, both of which are possibly biased due to an insufficient number of records and due to hydraulic and mechanical limitation of tide gauges. There are many sources of error. The initial profile, determined with seismic data, can even be different from the actual tsunami profile. The numerical scheme introduces errors. Nonlinearity near and on land requires an appropriate selection of equations. Taking these facts into account, it should be noted that numerical simulations produce satisfactory information for practical use, because the final error is usually within 15% as far as the maximum run-up height is concerned.

The state-of-the-art of tsunami numerical simulations is critically summarized from generation to run-up. Problems in the near future are also stated. Fruitful application of computer graphics is suggested.

Key words. Tsunami initial profile, tsunami propagation, tsunami run-up, tsunami numerical simulation, tsunami secondary disaster, computer graphics.

1. Introduction

Numerical simulation has made much progress during the past 30 years and is now used as one of the most effective means in the practical design of tsunami defense works and structures. This progress was supported by the development of seismology and by the appearance of the high-speed computer. Seismology made it possible to estimate the fault mechanism and the related displacement of the sea bottom. This gives the initial profile of a tsunami which has never been measured on site with conventional methods of measurement. A huge near-field tsunami such as the 1933 Showa Great Sanriku tsunami has an extension, one to two hundred kilometers long and several tens of kilometers wide, when it is generated. The coast to be affected is several hundred kilometers long. If the effect of a distant tsunami such as the 1960 Chilean tsunami is discussed, the whole Pacific Ocean should be covered by a net of computation grids. This situation is only solved with electronic computers which ensure a huge computer memory and high-speed computation. A numerical simulation, if designed without due consideration, cannot provide any reliable results. There are many sources of error and misjudgement. If well-designed, its results can be used in practical tsunami defense works with sufficient accuracy, within 15% error as far as the maximum run-up heights and the inundated areas are concerned. The aim of the present paper is to provide a critical review of the state-of-the-art of numerical simulation, to summarize problems which require further study in the near future, and to show utility of a new technique in understanding computed tsunamis.

In Section 2, the generation problem is examined, taking the 1983 Nihonkai-Chubu (the Middle Japan Sea) earthquake tsunami as an example. Contradictions are found among tsunami sources determined from seismic data, tide records, and run-up heights. Limitation of tide gauges is also discussed.

In Section 3, the propagation problem is discussed, mainly as the problem of fundamental equations and difference schemes. It is evident that a set of equations of higher-order approximation yield better results, if used with fine spatial grids. An ingenious method is introduced to obtain equivalent results with equations of first-order approximation and with rough spatial grids, thus saving much of the CPU time and computer money.

Section 4 reviews the near-shore and run-up problem, in which the major factors we encounter are the nonlinearity of the phenomena, instability in the computation, and approximation of the moving land boundary.

In Section 5, three topics are briefed for future study. A theory must be developed for edge bores which were observed in 1983. In relation to secondary disasters, spread of oils as well as spread and impact of floating materials are in urgent need of study. Computer-graphics-aided animation is promising as a way to understand tsunamis through computed results.

2. Generation of Tsunamis

2.1. *Three Source Models of the 1983 Nihonkai-Chubu Earthquake Tsunami*

At noon on the 26 May 1983, a huge earthquake occurred in the Japan Sea. A tsunami followed. Its highest run-up was measured at more than 15 m above the mean sea-water level. Several features of the tsunami were made clear by tide records, detailed surveys of run-up height and inundated areas, and many photos and videos.

As for the earthquake, more scientific means were available. For example, positions of the major shocks and every after-shock were three-dimensionally determined as in Figure 1 (Takagi *et al.*, 1984). An immediate conclusion is that the fault dips slightly eastward. Fault mechanism and fault parameters were determined from seismic data collected not only locally but also world-widely.

Figure 2(a) shows the vertical displacement of the sea bottom, calculated by Tanaka *et al.* (1984) with the Mansinha and Smylie (1971) method. Since the source was large compared with the water depth and the rupture velocity was very short compared with the tsunami propagation velocity, the bottom vertical displacement gave the vertical displacement of the free water surface, i.e. the initial tsunami profile. Only seismic data were used in this calculation. The major result is that a

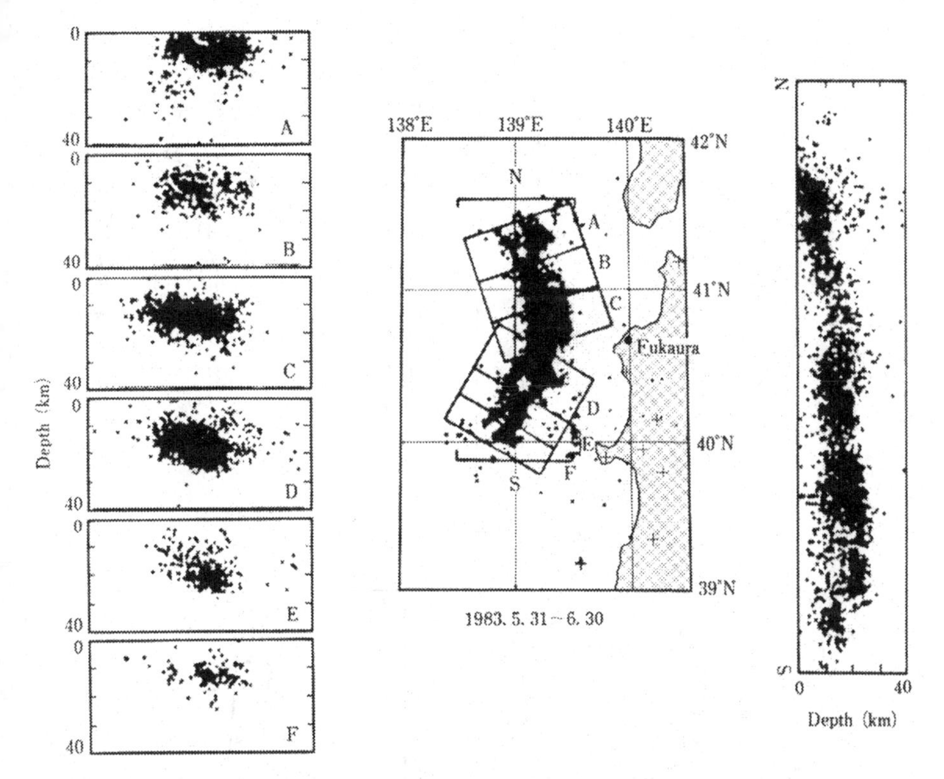

Fig. 1. After-shocks of the 1983 Nihonkai-Chubu earthquake.

maximum vertical displacement of about 1.5 m high is obtained at the western side of the source, corresponding to the fault which dips eastward. Tanaka *et al.* verified the computed displacement by comparison with the measured displacement at Kyuroku Island.

Figure 2(b) is the result obtained by Satake (1985). He used the tide records at 13 tide stations to modify the bottom displacement obtained from seismic data. The method of modification will be briefed in Section 2.3. According to his results, the fault dips eastward and the maximum vertical displacement is about 2 m.

Figure 2(c) is the result of a thorough study by Aida (1984a). The fault dips eastward and the maximum vertical displacement is about 4 m. He needed this maximum vertical displacement in order to explain run-up heights. He used seismic data, tide records at remote tide stations, and run-up data on the coast near the source. Based upon his rich experience, Aida did not use tide records at tide stations near the source, because he might be led to an incorrect conclusion by the fact that the hydraulic characteristics of a tide well often render it unable to record high frequency components of a nearby tsunami. He used run-up data in a way described in Section 2.3. His conclusion was verified by the present author who used

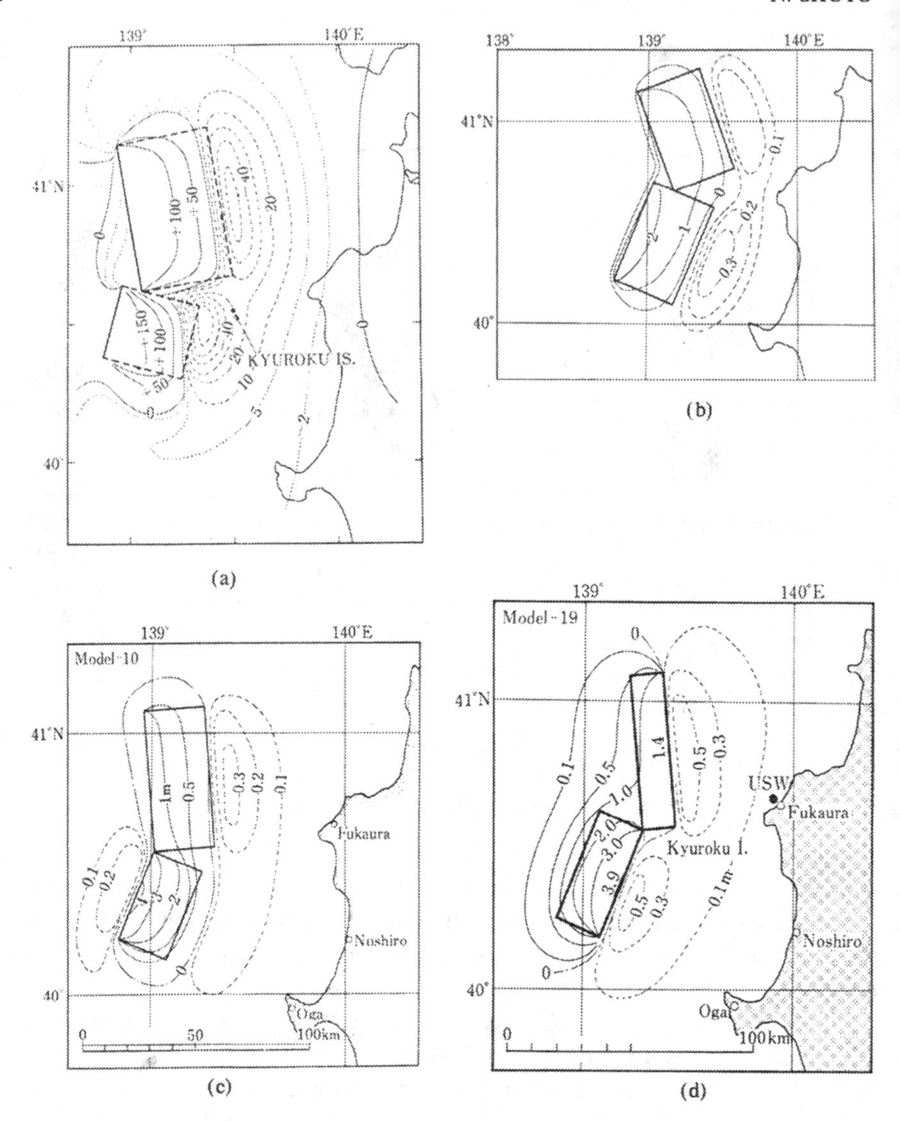

Fig. 2. Vertical displacement of sea bottom due to the 1983 Nihonkai-Chubu earthquake. (a) Tanaka *et al.*, (b) Satake Model D2, (c) Aida Model No. 10, and (d) Aida Model No. 19.

a finer grid length than usual and obtained the maximum run-up height of 15 m. The results has been published elsewhere (Shuto *et al.*, 1986).

2.2. *Aida Model No. 19, a Contradiction*

An ordinary tide gauge has three problems in recording tsunamis: hydraulic filtering, poor resolution with respect to time, and possible saturation in the case of a huge tsunami. A tide well is usually constructed to eliminate high-frequency sea-level fluctuation, such as wind waves and swells. This characteristic also deteriorates the tsunami recording ability of a tide gauge. Satake *et al.* (1988) measured the hydraulic filtering characteristics of 40 tide wells and corrected tsunami records of the 1983 Nihonkai-Chubu earthquake tsunami. They found that some tide records should be more than doubled in amplitude.

This results, and the poor temporal resolution, suggest that we should be very careful when we compare the computed tsunami profiles with tide records for verification of the initial tsunami profile. Knowing this fact well, Aida tried another effort to estimate the initial profile of the 1983 tsunami. It was fortunate for him that a supersonic wave gauge for measuring wind waves and swells was available. This instrument recorded the initial small fall and the succeeding sharp rise of the 1983 tsunami, although the signal was superposed by short wind waves. This wave gauge measured directly the vertical distance to the water surface without any filtering. The temporal resolution was excellent, because the wave gauge should measure wind waves of only several seconds long.

Aida thought this record gave the most reliable tsunami profile. He struggled again and again to reproduce this tsunami profile by numerical simulation and arrived at a surprising conclusion – the Aida Model No. 19, shown in Fig. 2(d) (Aida, 1984b). It required that the fault should dip westward, contradicting seismic knowledge, although the maximum vertical displacement of the sea bottom was about 4 m, the same as his Model No. 10.

A possible explanation to mediate the contradiction may be the existence of a secondary fault, as in the case of the 1964 Great Alaska earthquake, as shown in Figure 3 (Plafker, 1965).

In conclusion, the present method, based upon seismic data, can be applied to establish the initial tsunami profile to a first-order approximation, but it does not give a full explanation of the tsunami. In hindcasting, correction should be made by using measured data: tide records and run-up heights. In forecasting, there are no alternatives.

2.3. *Verification in Terms of Aida's K and κ*

In order to examine whether the assumed initial tsunami profile is satisfactory or not, Aida (1978) introduced two measures, K and κ. In his original paper, he compared the first rise or fall of computed profiles with measured tide records, with no regard to the succeeding waves which were much affected by local topographical effects. The measure K is a geometric mean of the ratio of the measured amplitude to the computed, and κ is the corresponding standard deviation. If K is smaller than unity, it means the assumed tsunami initial profile (or the assumed total tsunami energy) is larger than the true solution. Therefore, he could multiply the assumed initial vertical displacement by the obtained value of K to obtain a reduced estimate. Another measure κ shows whether or not distribution of the vertical displacement is appropriate.

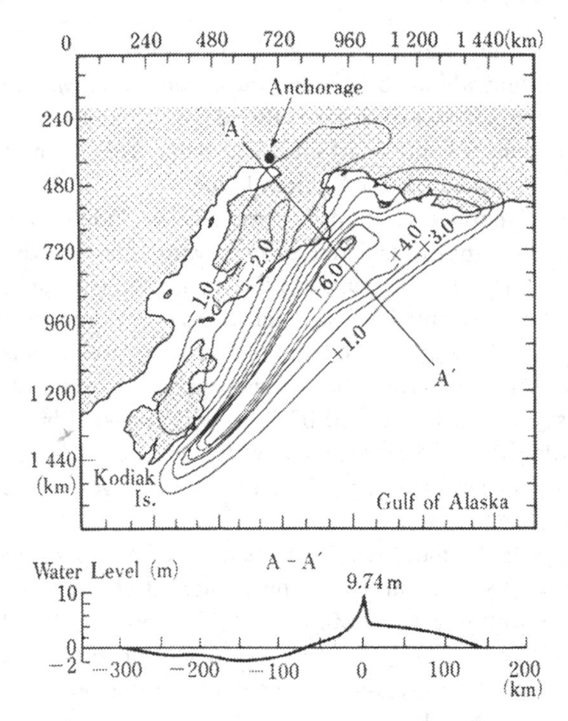

Fig. 3. Vertical displacement of ground and sea bottom of the 1964 Great Alaska earthquake.

Since the number of tide records available is usually 10 or so, not many for a tsunami, and since no tide record was available for a tsunami in the remote past, Aida extended the above method to apply to run-up heights, too. Measured run-up heights might strongly reflect local topography effects. He divided the coastal line into segments 10 or 15 km long. After smoothing the measured data by taking the average over each segment, he compared these smoothed values with the computed results obtained along the water depth contour of 200 m. If the ratios of the former to the latter are between 2 and 3 (= average run-up ratio), and if the value of κ is within an appropriate range, he judged that the assumed initial profile gave the solution.

With the initial profile thus given, a detailed numerical simulation can be carried out, including the reproduction of local topography effects. For the final judgement of whether or not the simulation gives satisfactory run-up heights, the same method is applied with no processing of the measured and computed data. For practical purposes, a numerical simulation is judged to be satisfactorily carried out if K falls between 1.2 and 0.8 and κ is less than 1.4 after necessary modification of the initial profile. Then, for a well-designed simulation, the final error is within 15% as far as the maximum run-up height is concerned (e.g., Aida and Hatori, 1984; Hasegawa, 1986; Shuto *et al.*, 1987).

3. Propagation in Deep Sea

3.1. *Linear Longwave Theory and Grid Length for No Decay*

At the time of generation, a tsunami in the water several kilometers deep is several tens of kilometers wide and up to 100 km long. Its height is less than 10 m. The depth-to-length ratio is on the order of 10^{-2} and the wave steepness is of order 10^{-3}. These values suggest that the linear longwave theory is a good first-order approximation.

Numerical simulations provide only approximate solutions, with errors inevitably included in and dependent on numerical techniques. In the following discussions, the leap-frog scheme is assumed for discretizing fundamental differential equations.

Figure 4 shows examples of computed wave profiles. Taking a sectional profile of the 1964 Great Alaska earthquake tsunami measured along the A–A line in Figure 3 as the initial profile, the one-dimensional propagation on water of constant depth is computed with the linear longwave theory. This theory gives a unique wave celerity which is not influenced by phase and amplitude dispersion effects. The true solution, therefore, should give the mere translocation without any change in the wave profile. It is evident, however, from the figures, that the wave profile deforms, depending upon the spatial grid length and the travel distance. The smaller the grid length is and the shorter the travel distance is, the truer the solution becomes. The change is that the leading wave reduces its height and a small wave-train appears behind it.

In order to eliminate this kind of numerical decay in wave height, the grid length should be carefully determined. According to numerical experiments by Shuto *et al.* (1986), one local tsunami wavelength should be covered by more than 20 grid points. Thus, the decay is less than 5% after the wave travels over a distance of four wavelengths, which is the longest travel distance for the first wave in the case of a typical near-field tsunami near the Japanese Archipelago. This condition should be satisfied, not only in deep oceans, but also in shallow seas.

3.2. *Dispersion Effect and Equations*

It was Kajiura (1963) who introduced a criterion to determine whether or not the dispersion effect should be taken into consideration. If his P_a defined by

$$P_a = (6h/R)^{1/3}(a/h) \tag{1}$$

is smaller than 4, the dispersion effect is not negligible. Here, h is the water depth, R is the travel distance, and a is the length of the tsunami source measured along the direction of propagation.

When the dispersion term is required, we have to switch from the linear longwave theory to the Boussinesq or linear Boussinesq equations. Under the same conditions as in Figure 4, the one-dimensional propagation is computed with the three equations: linear long wave, linear Boussinesq, and Boussinesq equations for two

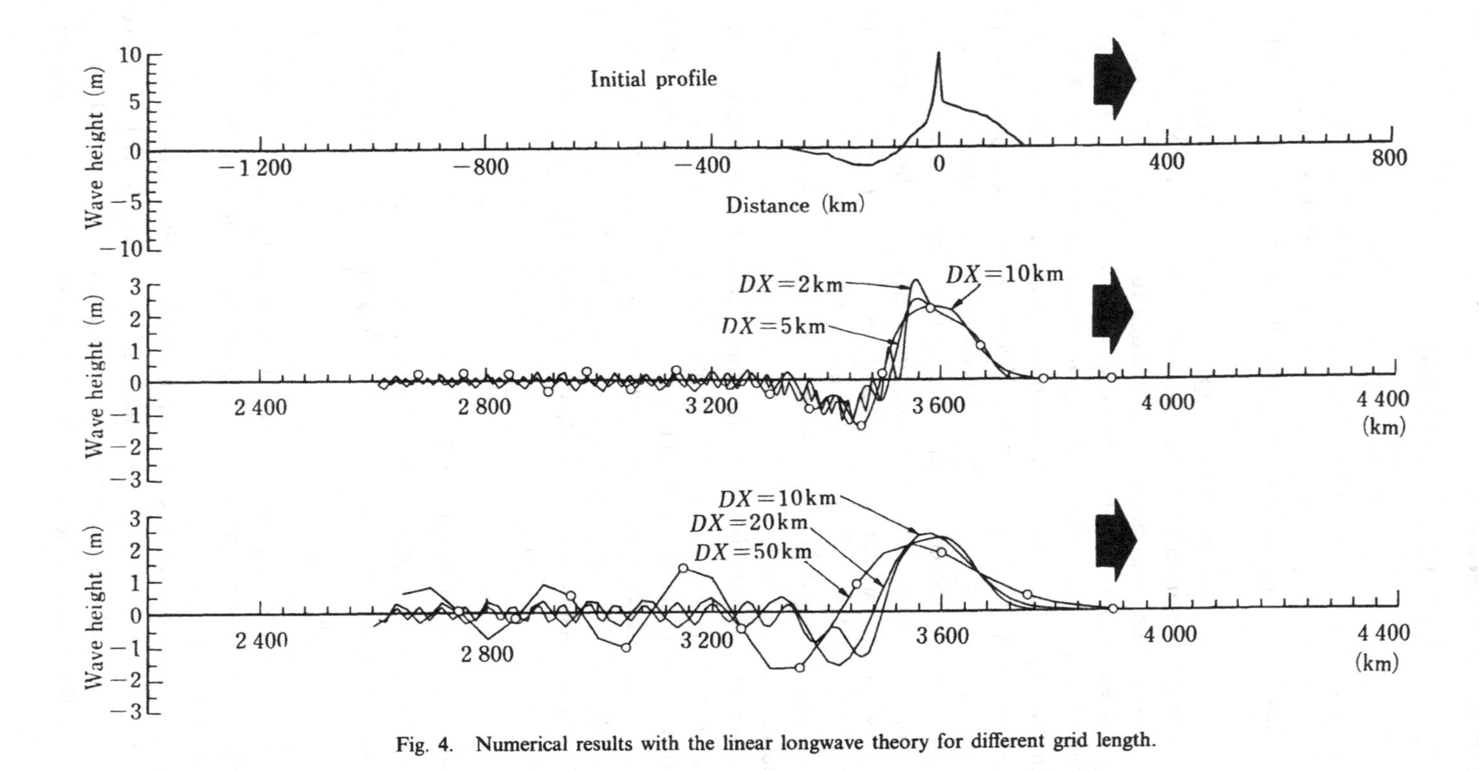

Fig. 4. Numerical results with the linear longwave theory for different grid length.

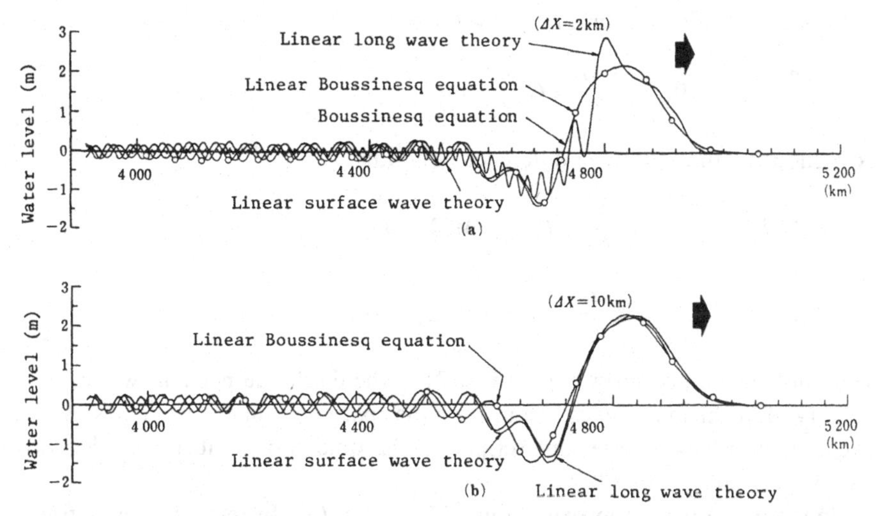

Fig. 5. Comparison of three longwave theories. The true solution is given by the linear surfacewave theory. (a) $\Delta x = 2$ km, and (b) $\Delta x = 10$ km.

different grid lengths. Figures 5(a) and 5(b) compare the results with the results of the linear surfacewave equation. By using the theory of linear surface waves, which fully includes the dispersion effect, the true solution is obtained as follows. The initial profile is decomposed into the Fourier components, which are translated with their own celerity. Then, the component waves are summed up to reconstruct the wave profile.

With the fine grid, the linear Boussinesq and Boussinesq equations almost coincide with the true solution, suggesting that the nonlinear term is not important in the propagation of a tsunami in deep ocean. The linear longwave theory deviates from the true solution in spite of the fine grid length.

On the other hand, with the coarse grid, the linear Boussinesq equation gives a satisfactory result for the major wave but a poor agreeement for the oscillation behind the major wave. This means that a higher-order equation with a fine grid gives better results. On the contrary, it is interesting to notice that the linear longwave equation with the coarse grid yields a better result. It is better than a higher-order equation with the same grid length and also better than itself with the fine grid length. This is due to the numerical dispersion.

3.3. *Numerical Dispersion and Its Use*

When a differential equation is reduced to a difference equation, approximation errors are inevitably introduced. The property and magnitude of the errors depend on the type of original equation and discretization scheme. In the following discussion, the staggered leap-frog scheme is used.

Equation (2) for the linear long waves,

$$\frac{\partial \eta}{\partial t} + \frac{\partial M}{\partial x} = 0, \qquad \frac{\partial M}{\partial t} + C_0^2 \frac{\partial \eta}{\partial x} = 0 \tag{2}$$

is reduced to the following difference equation

$$\begin{aligned} &\eta_{i+1/2}^{n+1/2} - \eta_{i+1/2}^{n-1/2} + \frac{\Delta t}{\Delta x}[M_{i+1}^n - M_i^n] = 0, \\ &M_i^{n+1} - M_i^n + C_0^2 \frac{\Delta t}{\Delta x}[\eta_{i+1/2}^{n+1/2} - \eta_{i-1/2}^{n+1/2}] = 0. \end{aligned} \tag{3}$$

Here, η is the water surface elevation, M is the discharge per unit width, C_0 is the celerity of linear long waves, x and t are the space and time coordinates, Δx and Δt are their grid lengths, the superscript n is the time grid number, and the subscript i is the spatial grid number.

Equation (3) is an approximation of Equation (2). In reverse, if Equation (3) is regarded as the true equation, Equation (2) is an approximation having errors. On making corrections to include the first term of the errors, a better differential expression of Equation (3) than Equation (2) is given as follows

$$\frac{\partial^2 \eta}{\partial t^2} - C_0^2 \frac{\partial^2 \eta}{\partial x^2} - \frac{C_0^2 \Delta x^2}{12}(1 - k^2)\frac{\partial^4 \eta}{\partial x^4} = 0, \tag{4}$$

in which k is the Courant number defined by $k = C_0 \Delta t / \Delta x$. The third term is the numerical dispersion. Although it is of the order of the square of the grid length, the accumulated effect becomes nonnegligible after a long travel.

The linear Boussinesq equation, a higher-order longwave equation which has the physically required dispersion term, is given by

$$\frac{\partial^2 \eta}{\partial t^2} - C_0^2 \frac{\partial^2 \eta}{\partial x^2} - \frac{C_0^2 h^2}{2} \frac{\partial^4 \eta}{\partial x^4} = 0, \tag{5}$$

in which h is the water depth.

It is Imamura (1989) who proposed the use of numerical dispersion in place of physical dispersion, by setting equal the coefficients of the third terms in Equation (4) and (5). When the grid length is selected so that the following Imamura number is nearly equal to unity

$$I_m = \Delta x[1 - (C_0 \Delta t / \Delta x)^2]^{1/2} / 2h, \tag{6}$$

the linear longwave equation gives the same results as the linear Boussinesq equation, thereby saving much of the CPU time and computer memory.

Since the dispersion effect is not a strong effect, it is not necessary to make the Imamura number strictly equal to unity. Figure 5(b), in which the Imamura number is 1.4, shows a good agreement between the true solution and the result of the linear longwave theory.

3.4. *Coriolis Force*

In a simulation of a tsunami traveling the Pacific Ocean, the Coriolis force is necessarily included. However, in the 'ray method' often used to forecast the arrival time of a tsunami, no Coriolis force is taken into consideration. Figures 6(a) and 6(b) compare the contours of the highest water level in the Pacific Ocean computed for the 1969 Chilean tsunami with and without the Coriolis force. Due to the Coriolis force, the divergence and the convergence of the tsunami, i.e. its height as a result, are different, but the arrival time shows no large difference.

4. Near-Shore and Run-Up Problems

4.1. *Nonlinearity and Equations*

Entering shallow water and approaching the shore, a tsunami increases in height, steepness, and curvature of water surface. Equations change according to the order of approximation required to describe it: the linear longwave theory, the shallow-water theory, the Peregrine equation (1967) and the Goto equation (1984).

In seas deeper than 50 m, the linear longwave theory gives satisfactory results. As water depth decreases, the equations should be switched to the shallow-water theory with bottom friction. Generally speaking, the shallow-water theory is believed sufficient in tsunami simulation as far as the maximum run-up height and inundated area are concerned. It is, however, not sufficient in regard to the estimated wave force, which is closely related to wave profiles.

Figure 7 compares two higher-order equations with hydraulic experiments. The Peregrine equation shows an early dispersion which accelerates the fission of waves and leads to an increase in height. Although the Goto equation gives better results in this example, the range of its application has not yet been examined. In addition, if a higher-order equation is extended to the two-dimensional problem, the computation becomes extremely complicated. No attempt has been made to carry out a two-dimensional tsunami simulation with equations higher than the shallow-water theory.

4.2. *Examples of Instability*

4.2.1. *Instability near the Side Boundary.* When the capacity of a computer is not large enough compared with the number of grid points and the time required for a full computation, it is a usual technique to divide the computation into two stages, the offshore computation and the near-shore computation.

In the offshore computation, the entire region, from the tsunami source area to the land boundary, is covered with a net of coarse grids 3 and 5 km long for a near-field tsunami and about 10 km long for a far-field tsunami. The offshore computation assumes perfect reflection from the shoreline, the shape of which is

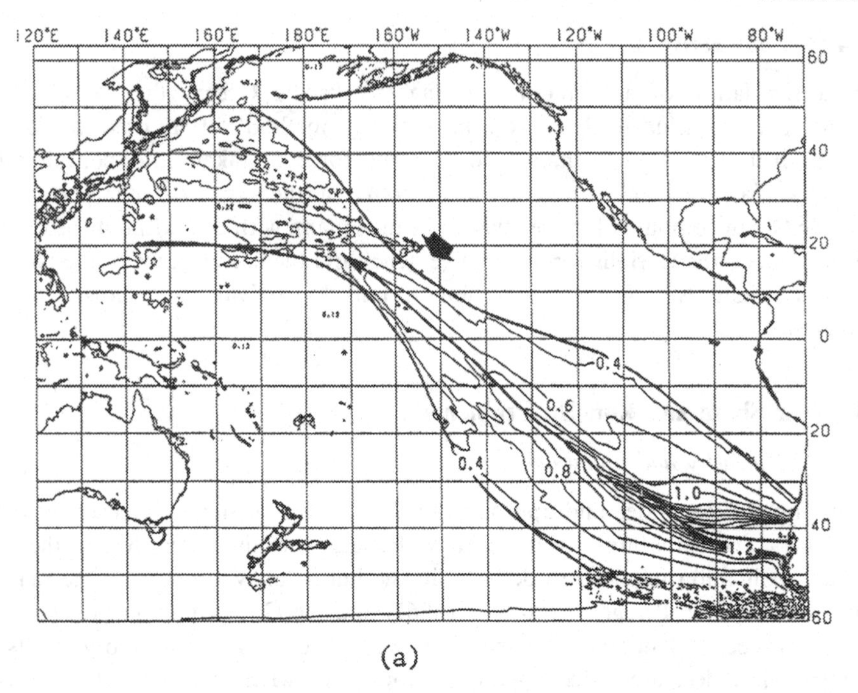

(a)

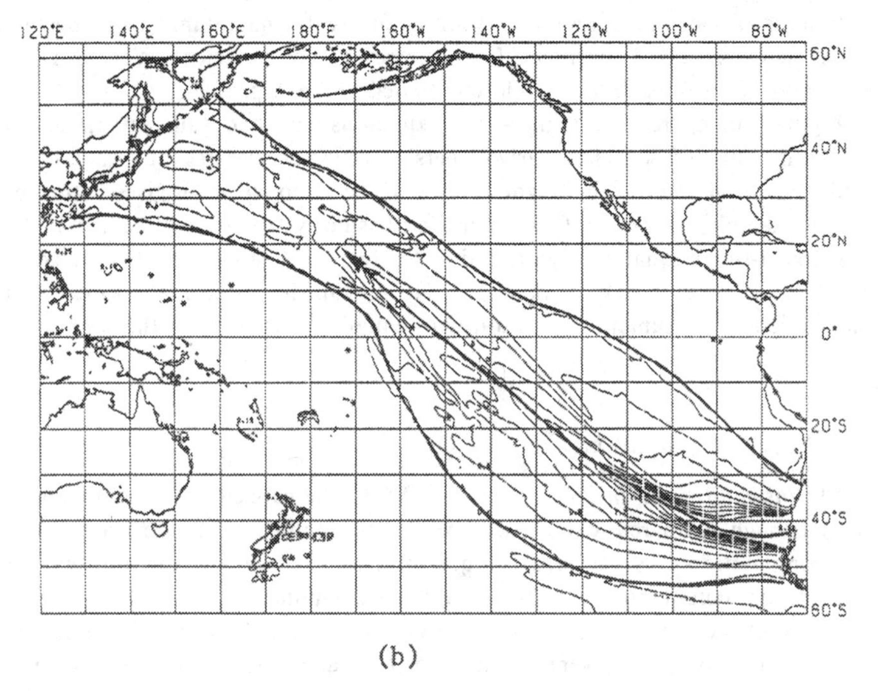

(b)

Fig. 6. Computed highest water level of the 1960 Chilean tsunami (a) with and (b) without the Coriolis force.

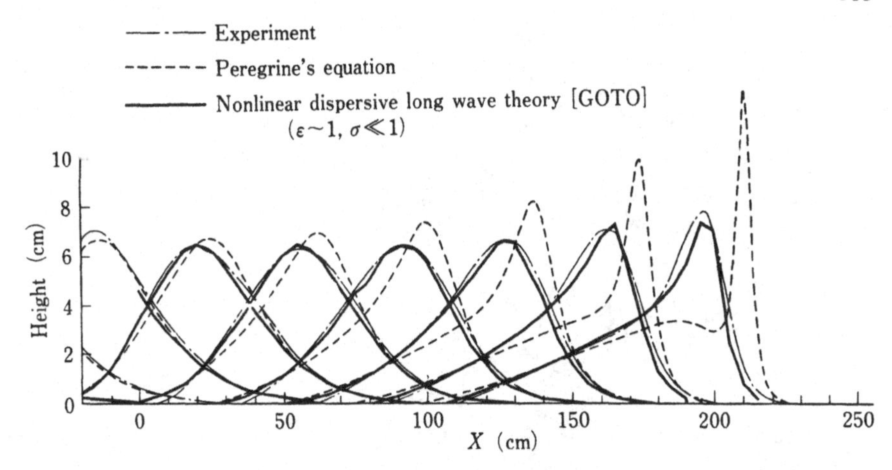

Fig. 7. Computed results with the Peregrine and Goto equations compared with hydraulic experiment.

very roughly approximated. Computed results, therefore, reflect effects of topography on a large scale. In other words, the results are only reliable at points in deep sea far from the land boundary, where effects of run-up and small-scale topography become unimportant.

The next stage is the near-shore computation, in which a net of fine grids varying from 1 km in deep sea to several tens of meters on land is used to allow observation of more limited areas, including areas of special concern. The area in the nearshore computation is enclosed by four boundaries: the land boundary, the open-sea boundary (which is usually set in the neighborhood of 1000 m contour and nearly parallel to the land boundary), and two side boundaries normal to the open-sea boundary. When outputs of the offshore computation at the coarser grid points are inputted on the open-sea and side boundaries in the near-shore computation, linear interpolations are used to determine the input at the finer grid points interposed between the coarser grid points. These interpolated values are only approximations, not the true solution, which will be obtained with the near-shore computation. Before waves reflected from land arrive at the boundaries, no difference between the assumed and true boundary values give serious influence to the computation. Even when reflected waves arrive at the far boundaries in deep sea, no serious discrepancy is found between the assumed input and the results of the detailed near-shore computation, because wave components reflecting effects of small-scale topography rapidly die out. On the contrary, in the shallow sea, in particular in the neighborhood of a corner where the major shoreline and the side boundary intersects with an acute angle, the reflection immediately arrives at the side boundary and often gives a considerable difference between the interpolated values and the computed results. This is the cause of an instability which will propagate rapidly and demolish the whole computation. Figure 8 shows an example of this kind of instability.

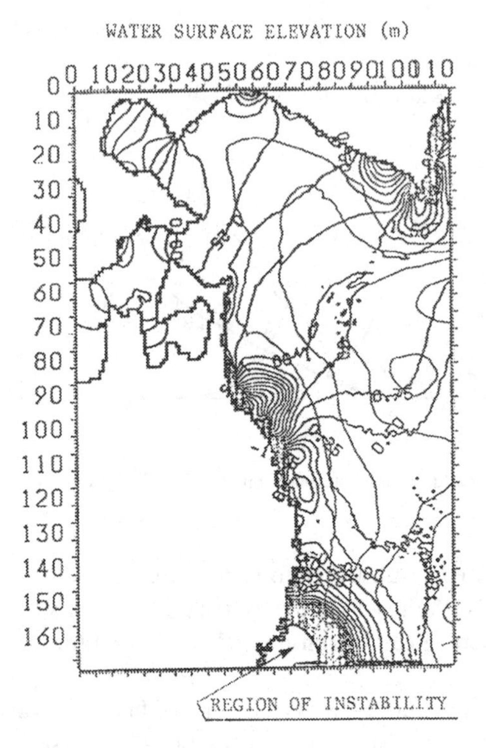

Fig. 8. Initiation of an instability at a corner of the boundary.

4.2.2. *Instability at the Front.* Figures 9(a) and 9(b) show examples of another instability. This oscillation occurs at the front of the second wave, which is running up against the back wash of the first wave. The second wave is retarded, its front surface steepens, and the oscillation occurs. With a smaller grid length, the length of instability waves is shorter. Evidently, this is a numerical instability dependent on the grid length used.

A method to eliminate this instability was introduced by Goto and Shuto (1983a). They use an artificial diffusion which acts to cancel the instability waves only in the vicinity of the wave front. When the artificial diffusion acts well, the front surface becomes more gently sloping than the front should be. An artificial viscosity is used to amend this over-smoothing, at the expense of negligible dissipation of energy. Figure 9(c) is an example of this method.

4.3. *Moving Boundary Approximation*

A tsunami front runs up and down the land. It is not easy to express this moving boundary with the coordinate system in the Eulerian description. If equations in the Lagrangian description are used, the moving boundary condition can be expressed

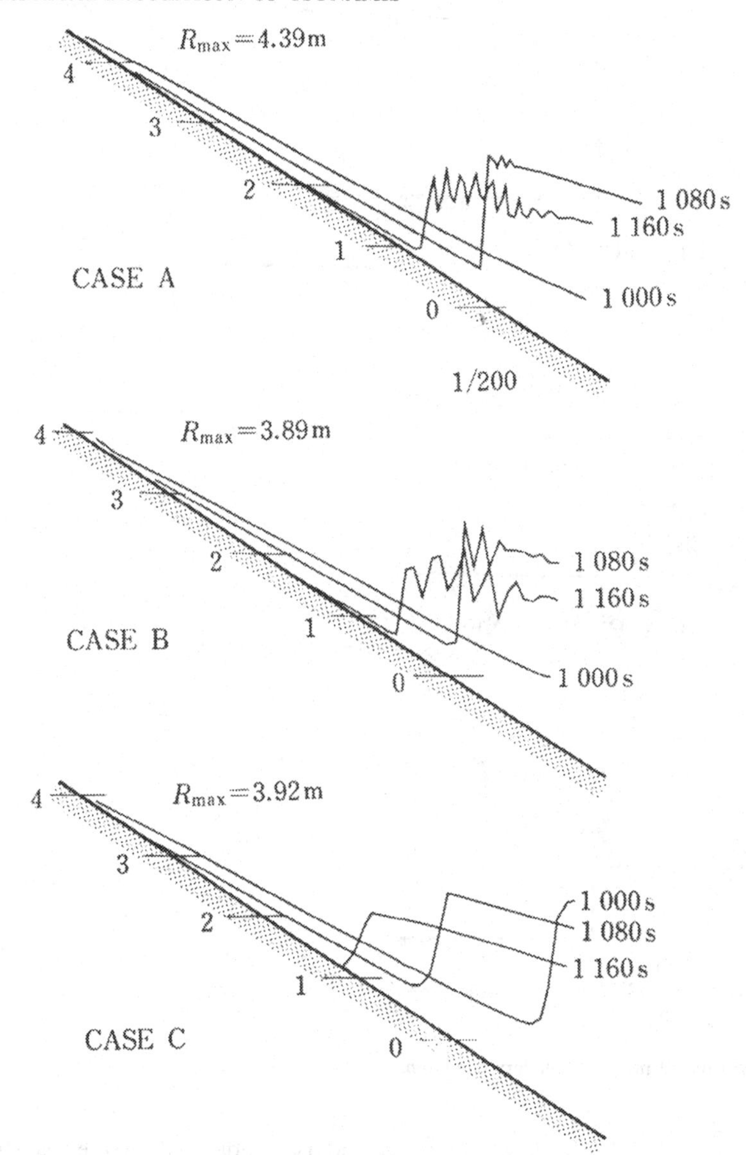

Fig. 9. Oscillation induced at the wave front (a) with $\Delta x = 12.5$ m and (b) with $\Delta x = 25$ m. (c) Wave profiles without oscillations obtained using the artificial diffusion term.

with no approximation (Shuto and Goto, 1978). Even with equations in the Eulerian description, a variable transformation in which the origin of the new coordinates is located at the front can easily express the moving boundary (Takeda, 1984). It is unfortunate, however, that these two methods are well applicable only to one-dimensional problems, but poorly to any two-dimensional practical problem with complicated topography.

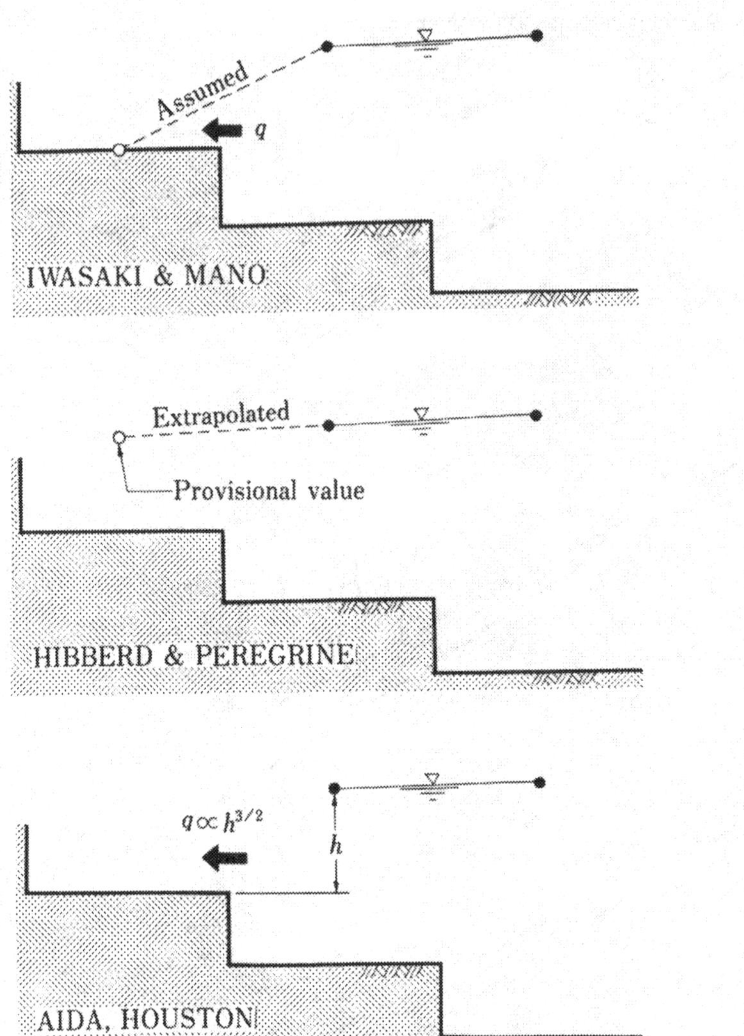

Fig. 10. Approximated moving boundary conditions.

There are several approximate moving boundary conditions. In the leap-frog scheme, grid points are alternatively located for velocity and water level. Assume that the water level is already computed at a computation cell. Then, compare the water level with the bottom height of the next landward cell. If the water level is higher than the latter, the water may flow into the landward cell. Figure 10 explains the ways to estimate the inflow velocity or discharge.

Iwasaki and Mano (1979) assume that the line connecting the water level and the bottom height gives the surface slope to the first-order approximation.

Hibberd and Peregrine (1979) give a provisional water level in the dry cell on a linearly extrapolated water surface. Then, the discharge calculated with this provisional water level gives the total amount of water into the dry cell and the water depth in the cell. If necessary, the computation will be repeated with the water level thus modified.

Aida (1977) and Houston and Butler (1979) evaluate the discharge into the dry cell with broad-crested weir formulas in which the water depth above the bottom of the dry cell is substituted.

These approximations are convenient to handle but introduce numerical errors (Goto and Shuto, 1983a). The run-up height computed with the Iwasaki–Mano method agrees with the theoretical solution with a 5% range of error if the following condition is satisfied

$$\Delta x / \alpha g T^2 < 4 \times 10^{-4}. \quad (7)$$

With the Aida method, the condition is given by

$$\Delta x / \alpha g T^2 < 10^{-3}, \quad (8)$$

in which α is the angle of slope, g is the gravitational acceleration, Δx is the spatial grid length, and T is the wave period.

In closing this section, a comment should be made on the widely-believed opinion that there is no difference between linear and shallow-water theories as far as the maximum run-up height is concerned. In the case of one-dimensional problems, the maximum run-ups are the same even though the two theories give different wave profiles. On the other hand, in a practical problem in which the land has a two-dimensional complicated topography, the lateral flow is induced and affected much by the difference in wave profile; this nonlinearity becomes important.

4.4. *Effect of Large Obstacles*

There are three methods to include the effects of building into tsunami simulation. The simple method in a hindcasting is to allot a large friction coefficient f, which ranges from 0.2 to 1, to the residential areas. This method is, however, not applicable to a forecasting, because an appropriate value of the friction coefficient is determined only after comparisons of the computed inundated area with the recorded. A more reasonable method which can be used in a forecasting is to determine an equivalent friction coefficient by summing up the drag of individual buildings (Goto and Shuto, 1983b). The best way is to use very fine grids in the city area. If the grids are less than 5 m wide, most of the large buildings can be expressed as impermeable boundaries. This inevitably increases the number of grid points. Fine grids also introduce the question of whether or not a map used in discretization is accurate enough for this detailed computation.

In the neighborhood of a large obstacle covered by fine grids, the water flow can be numerically simulated. However, this leaves the question of whether the computed result is reliable. Figure 11 shows an example. Uda *et al.* (1988a) computed a tsunami that overflowed a model sand dune. They compared the results with the

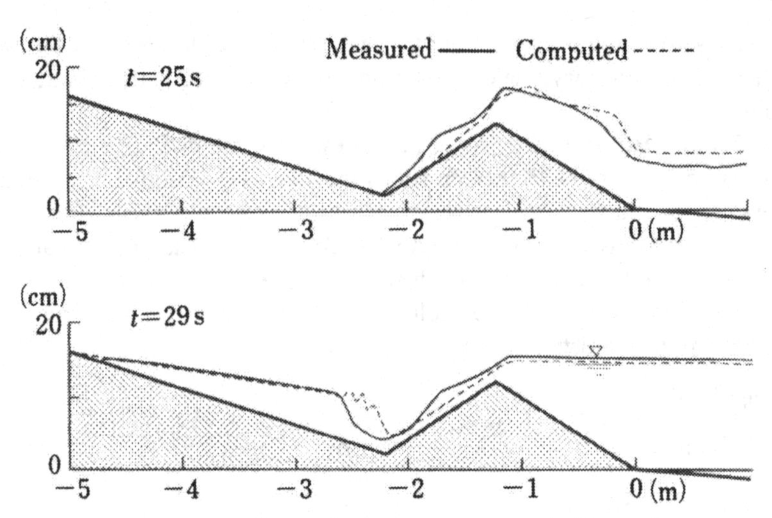

Fig. 11. Computed tsunami profiles overflowing a dune compared with experiments.

measured data in hydraulic experiment. Agreement was poor just behind the dune where the water flow varied from supercritical to subcritical flows through a jump. Then, agreement recovered further behind the dune.

5. Topics for Future Study

5.1. *Edge Bore*

After entering the North Akita coast 55 km long, the 1983 Nihonkai-Chubu earthquake tsunami was trapped along the slightly curved coast of a gentle sandy beach. In photos taken from a coastal hill, it was shown that edge bores hit the coast again and again. Hydraulic experiments carried out in the Public Works Research Institute (Uda *et al.*, 1988b) revealed several interesting phenomena. An edge bore propagates sometimes following the ordinary refraction law and sometimes neglecting the topography. A small difference in the boundary condition might introduce a big difference in wave profile, thus suggesting also a big difference in wave force. No theory and no simulation method are available now.

5.2. *Spread of the Floating Materials*

In the coastal areas, large quantities of oils and timbers are often stored and small fishing boats are often moored in the vicinity of harbors. The presence of these materials increases the danger to the surrounding areas if transported by a tsunami. In 1964, three towns in Alaska were heavily damaged by fires due to oil spills spread by the tsunami. On the occasions of huge tsunamis in the past, timbers, boats, and debris of broken houses changed into formidable destructive forces.

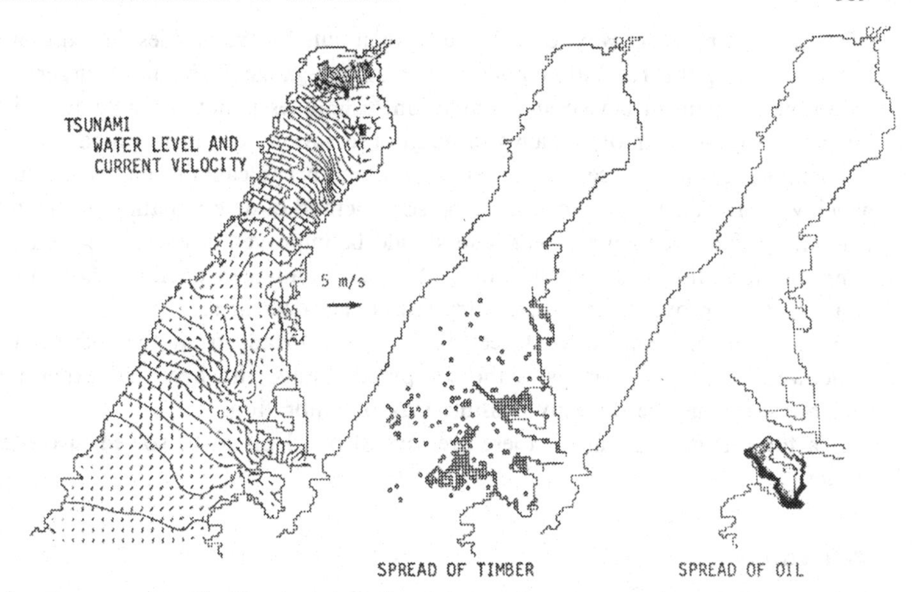

Fig. 12. Spread of oil and timbers due to a tsunami.

Goto developed numerical methods to estimate the spread of oil (1985) and timbers (1983). Figure 12 shows an example. However, these methods have not yet been checked with experimental or measured data.

Although the impact of timbers needs an urgent study for practical design, there is only one paper (Matsutomi, 1989) on this problem.

5.3. *Computer Graphics*

A tsunami numerical simulation creates huge quantities of information, only a few of which are used at present. They are the maximum run-up along the coast, the inundation area, the time history of water level at several selected points, velocity vectors and water level contours at several time intervals, and so on. If a computer-graphic-aided video animation is introduced, the whole computed results would be efficiently used and would reveal the detailed dynamic movement of the tsunami from several viewpoints. The video animation may replace the hydraulic experiment on a large scale to some extent, if we keep in mind that the motion in the animation is limited by the theory used in the simulation.

6. Concluding Remarks

The numerical simulation is now used as a powerful means in the planning of tsunami defense works. A hindcasting can give the maximum run-up height and the corresponding flooded area within a 15% error. This does not, however, mean that tsunami waves are always correctly reproduced. From the generation to the final effect, there is no exact way to calibrate the computed wave profiles, because of the shortage and poor resolution ability of tsunami gauges. The most urgent and necessary problem in developing simulation techniques is not the simulation itself but revealing the truth of tsunamis through observation networks of high accuracy. Tsunami gauges in deep sea are not enough in number to catch tsunamis in infancy, which will provide a way to check the present method of determination of the initial tsunami profile. Tide gauges near land should be improved to overcome their poor temporal resolution and possible scale-out in case of a huge tsunami, although there is a way to improve its hydraulic attenuation characteristics.

As for currents and wave forces that are important from the viewpoint of structure design, comparisons of the computed results with hydraulic experiments of large scale may be the only possible way of calibration.

If a forecasting is required, there are several problems to be solved in the near future.

References

Aida, I.: 1977, Numerical experiments for inundation of tsunamis, Susaki and Usa, in the Kochi Prefecture, *Bull. Earthq. Res. Inst.* **52**, 441–160 (in Japanese).

Aida, I.: 1978, Reliability of a tsunami source model derived from fault parameters, *J. Phys. Earth* **26**, 57–73.

Aida, I.: 1984a, A source model of the tsunami accompanying the 1983 Nihonkai-Chubu earthquake, *Bull. Earthq. Res. Inst.* **59**, 235–265 (in Japanese).

Aida, I.: 1984b, Source models of the 1983 Nihonkai-Chubu earthquake tsunami, *Proc. Symp. Nihonkai-Chubu Earthquake Tsunami*, JSCE, pp. 9–21 (in Japanese).

Aida, I. and Hatori, T.: 1983, Numerical simulation of tsunami inundation in Yuasa and Hirokawa towns, Wakayama Prefecture in southwestern Japan, *Bull. Earthq. Res. Inst.* **58**, 667–681 (in Japanese).

Goto, C.: 1983, Numerical simulation of spread of floating timbers, *Proc. 30th Conf. Coastal Eng.*, JSCE, pp. 594–597 (in Japanese).

Goto, C.: 1984, Equations of nonlinear dispersive long waves for a large Ursell number, *Proc. JSCE* **351/II-2**, 193–201 (in Japanese).

Goto, C.: 1985, A simulation model of oil spread due to tsunamis, *Proc. JSCE* **357/II-3**, 217–223 (in Japanese).

Goto, C. and Shuto, N.: 1983a, Numerical simulation of tsunami propagations and run-up, in K. Iida and T. Iwasaki (eds.), *Tsunamis: Their Science and Engineering*, Terra Science Pub. Co., Tokyo/Reidel, Dordrecht, pp. 439–451.

Goto, C. and Shuto, N.: 1983b, Effects of large obstacles on tsunami inundations, in K. Iida and T. Iwasaki (eds.), *Tsunamis: Their Science and Engineering*, Terra Science Pub. Co., Tokyo/Reidel, Dordrecht, pp. 551–525.

Hasegawa, K.: 1986, Numerical simulations to substantiate past tsunamis for the planning of prevention works, Doctoral dissertation, Kyoto Univ. (in Japanese).

Hibberd, S. and Peregrine, D. H.: 1979, Surf and run-up on a beach: a uniform bore, *J. Fluid Mech.* **95**, 323–345.

Houston, J. R. and Butler, H. L.: 1979, A numerical model for tsunami inundation, WES Tech. Rep. HL-79-2.

Imamura, F.: 1989, Possibility of tsunami numerical forecasting, Doctoral dissertation, Tohoku Univ. (in Japanese).

Iwasaki, T. and Mano, A.: 1979, Two-dimensional numerical simulation of tsunami run-ups in the Eulerian description, *Proc. 26th Conf. Coastal Eng.*, JSCE, pp. 70–74 (in Japanese).

Kajiura, K.: 1963, The leading wave of a tsunami, *Bull. Earthq. Res. Inst.* **41**, 535–571.

Mansinha, L. and Smylie, D. E.: 1971, The displacement of the earthquake fault model, *Bull. Seismol. Soc. Amer.* **61**, 1433–1400.

Matsutomi, H.: 1989, Impact of breaking bore accompanying floating timbers, *Proc. Coastal Eng*, JSCE, 36, pp. 574–578 (in Japanese).

Peregrine, D. H.: 1967, Long waves on a beach, *J. Fluid Mech.* **27**, 815–827.

Plafker, G.: 1965, Tectonic deformation associated with the 1964 Alaska earthquake, *Science* **148**, 1675–1687.

Satake, K.: 1985, The mechanism of the 1983 Japan Sea earthquake as inferred from long period surface waves and tsunamis, *Phys. Earth Planet. Interiors* **37**, 249–260.

Satake, K. *et al.*: 1988, Tide gauge response to tsunamis: Measurements at 40 tide gauge stations in Japan, *J. Marine Res.* **46**, 557–571.

Shuto, N. and Goto, C.: 1978, Numerical simulation of tsunami run-up, *Coastal Engineering in Japan* **21**, 13–20.

Shuto, N. *et al.*: 1986, A study of numerical techniques on the tsunami propagation and run-up, *Sci. Tsunami Hazard* **4**, 111–124.

Shuto, N. *et al.*: 1987, Numerical simulation of tsunami propagations and run-ups on historical tsunamis—Summary, *Proc. Interational Tsunami Sym.*, NOAA, 184–187.

Takagi, A. *et al.*: 1984, Earthquake activities before and after the main shock, in K. Noritomi (ed.), General report on the disasters caused by the 1983 Nihonkai-Chubu earthquake, Rep. No. 58022002, supported by the Ministry of Education, Culture and Science, pp. 24–30 (in Japanese).

Takeda, H.: 1984, Numerical simulation of run-up by variable transformation, *J. Oceanogr. Soc. Japan* **40**, 271–278.

Tanaka, K. *et al.*: 1984, Characteristics of the Nihonkai-Chubu Earthquake, in K. Noritomi (ed.), General Rep. on the Disasters Caused by the 1983 Nihonkai-Chubu Earthquake, Rep. No. 58022002, supported by the Ministry of Education, Culture and Science, pp. 39–45 (in Japanese).

Uda, T. *et al.*: 1988a, Numerical simulation and experiment on tsunami run-up, *Coastal Eng. in Japan*, JSCE **31**, 87–104.

Uda, T. *et al.*: 1988b, Two-dimensional deformation of nonlinear long waves on a beach, Rep. Public Works Res. Inst., Minsistry of Construction, Japan, No. 2627 (in Japanese).

Natural Hazards 4: 193–208, 1991.

Use of Tsunami Waveforms for Earthquake Source Study

KENJI SATAKE and HIROO KANAMORI
Seismological Laboratory, California Institute of Technology, Pasadena, CA 91125, U.S.A.

(Received: 17 November 1989; revised: 25 May 1990)

Abstract. Tsunami waveforms recorded on tide gauges, like seismic waves recorded on seismograms, can be used to study earthquake source processes. The tsunami propagation can be accurately evaluated, since bathymetry is much better known than seismic velocity structure in the Earth. Using waveform inversion techniques, we can estimate the spatial distribution of coseismic slip on the fault plane from tsunami waveforms. This method has been applied to several earthquakes around Japan. Two recent earthquakes, the 1968 Tokachi-oki and 1983 Japan Sea earthquakes, are examined for calibration purposes. Both events show nonuniform slip distributions very similar to those obtained from seismic wave analyses. The use of tsunami waveforms is more useful for the study of unusual or old earthquakes. The 1984 Torishima earthquake caused unusually large tsunamis for its earthquake size. Waveform modeling of this event shows that part of the abnormal size of this tsunami is due to the propagation effect along the shallow ridge system. For old earthquakes, many tide gauge records exist with quality comparable to modern records, while there are only a few good quality seismic records. The 1944 Tonankai and 1946 Nankaido earthquakes are examined as examples of old events, and slip distributions are obtained. Such estimates are possible only using tsunami records. Since tide-gauge records are available as far back as the 1850s, use of them will provide unique and important information on long-term global seismicity.

Key words. Tsunami generation, tsunami waveforms, earthquake sources.

1. Introduction

Tsunamis are caused by various geological processes such as earthquakes, landslides, or volcanic eruptions. Among them, those caused by earthquakes are most frequent. Tsunami waveforms can be used to study earthquake sources in a way similar to that in which seismic waves are used. Most tsunami data used so far are arrival times and maximum heights. For example, the tsunami source area is usually estimated from tsunami arrival times using an inverse refraction diagram (e.g., Hatori, 1969). Maximum amplitudes can be used to estimate the water height in the source area using Green's law (e.g., Murty, 1977) or to calculate tsunami magnitude, M_t (Abe, 1979, 1989). We will use tsunami waveforms recorded on tide gauges to determine complex earthquake fault motion. This is a typical inverse problem to which waveform inversion techniques, extensively used in seismology, can be readily applied.

Figure 1 schematically compares the generation, propagation, and observation of seismic and tsunami waves. Both observed seismic and tsunami waveforms contain information not only on the source but also on the propagation paths and the

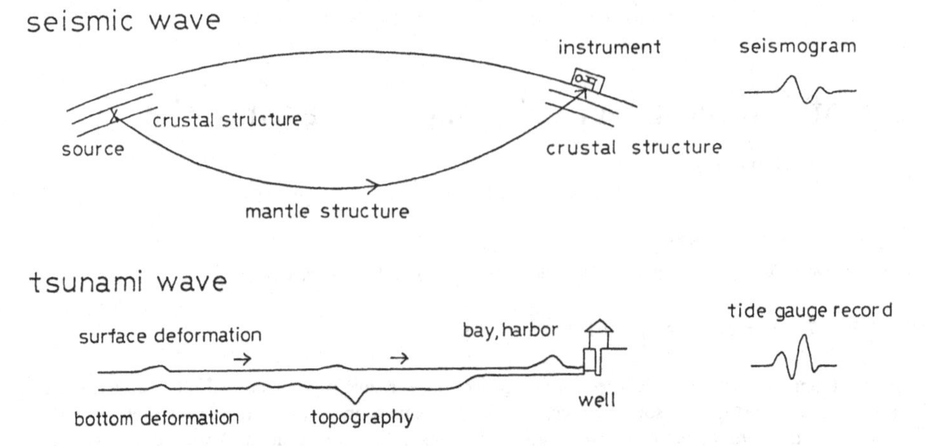

Fig. 1. Schematic illustration comparing seismic and tsunamic waves. Both seismograms and tide-gauge records contain information on the source.

instrumental responses. In the case of seismic waves, it is sometimes difficult to distinguish propagation and source effects, because the seismic velocity structure within the Earth is not completely known. For tsunami waves, on the other hand, the propagation effect can be easily computed numerically using actual bathymetry, because the bathymetry is known very well. In the following two sections, we will discuss the effects of propagation and the tide-gauge response. If we can remove these effects from the observed tsunami waveforms, the only remaining unknown is the source. We will formulate the inversion scheme and show examples in later sections.

2. Tsunami Propagation

There are basically two ways to treat tsunami generation and propagation, depending on whether the coupling of the water and the elastic Earth is considered or not. Some theoretical works made for the coupled system (Comer, 1984; Okal, 1988) showed that the coupling is negligibly small so that we can treat the water and the elastic Earth separately. In the uncoupled system, the deformation on the free surface of the elastic Earth caused by seismic sources is computed first. Then the deformed ocean bottom is considered as a rigid boundary for hydrodynamic computation of tsunamis. There are again two ways for hydrodynamic computation; one is analytic calculation on simple, mostly uniform, bathymetry and the other is numerical computation on realistic or actual bathymetry. The latter computation is becoming more popular because of development of the computer technology.

In hydrodynamic computation of tsunamis, two scale parameters are important to choose an appropriate equation: wavelength and amplitude. The tsunami wavelength is basically determined by the source size. For an earthquake with a magnitude $M > 7$, the source area has dimensions of several tens to hundreds of km, which is much larger than the ocean depth, at most 10 km. Thus, the tsunami wavelength from large earthquakes is longer than the water depth, ensuring the use of the nondispersive long wave (or shallow water) equation. However, the dispersion may not be negligible for tsunamis traveling a long distance such as the Pacific Ocean. We use tsunami waveforms recorded on tide gauges, that are usually installed at harbors where typical water depth is a few tens of meters. Since tsunami amplitude on gauges is usually up to a meter, small compared to the water depth, we can use small amplitude or linear theory. When tsunami run-up on beaches, where the water depth eventually becomes zero, is discussed, the nonlinear effect is obviously important. From the above discussion, the nondispersive, linear longwave equation is appropriate for the present purpose. The phase velocity of the nondispersive linear long waves is simply given by $\sqrt{gd}$ where d is the water depth.

The effect of bathymetry on tsunami propagation can be seen in the ray-tracing experiment of tsunamis across the Pacific Ocean (Figure 2). The panels on the left in the figure are the ray tracing for actual bathymetry, whereas the panels on the right are for an ocean of uniform depth. Focusing and defocusing of rays are caused by both geometrical spreading and lateral variation of tsunami velocity for the actual ocean, while only geometrical spreading can be seen in the uniformly deep ocean. For the Chilean source, rays are refracted at the East Pacific Rise and converged near Japan. In fact, tsunamis from the 1960 Chilean earthquake were as large as 8 m on the Japanese coast and caused more than 100 casualties. Focusing and defocusing of rays are also seen for the Aleutian source. The Japanese source has ray concentration near Chile, reciprocal to the Chilean source. The ray tracings of tsunamis show that bathymetric effects are significant on tsunami propagation. Actually, tsunami velocity structure is too heterogeneous to use the ray-theoretical approach.

For such a heterogeneous structure, the wave-theoretical approach must be used. We use a finite-difference method and compute the total tsunami wave-field. Basic equations are the nondispersive linear longwave equation and the equation of continuity. These equations are solved on a staggered grid system. We use a variable grid size system; coarser grids in the outer sea, but finer grids near coast especially near the tide-gauge stations. The details of the computation is given in Satake (1989).

3. Tide-Gauge Responses to Tsunamis

Tsunamis waveforms are usually recorded on a tide gauge which was originally designed to monitor ocean tides. The response of a tide-gauge system to tsunamis has not been well examined until 1983 when tsunamis from the Japan Sea earthquake were visually observed at many places in Japan. At some places, the visually observed tsunami heights were significantly larger (more than double) than

Fig. 2. Ray tracing of tsunamis in the Pacific Ocean. The panels on the left are for actual bathymetry while the panels on the right are for uniform depth (Satake, 1988).

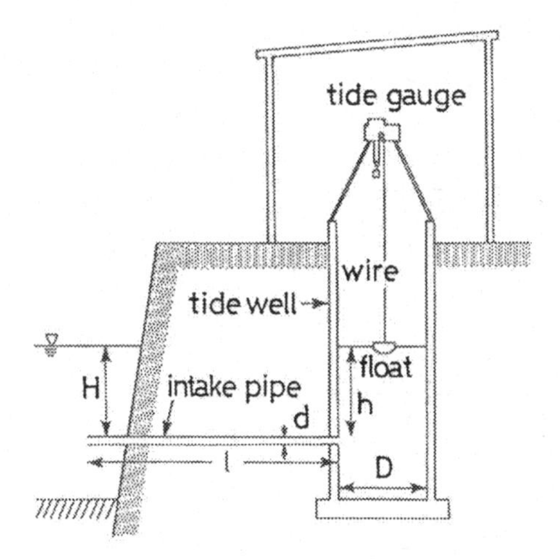

Fig. 3. Typical structure of Japanese tide-gauge stations (Satake *et al.*, 1988).

the tsunami amplitudes on the nearby tide-gauge record. The discrepancy is attributed to tide-gauge response to tsunamis.

Japanese tide-gauge stations have a typical structure as shown in Figure 3; the gauge is set in a well which is connected to the outer sea by an intake pipe. Because of the severe winter weather in the Japan Sea, some stations on the Japan Sea coast have designs to filter out the shorter-period components compared to ocean tides, such as surf or ocean beat, from tide-gauge records. Because the period of tsunami waves is shorter than that of ocean tides, tsunamis are also affected by the structure. The response of the tide-gauge system can be measured in a way similar to the impulse response measurement of seismographs. By putting water into the well or taking it out from the well, an artificial water-level difference is created and the following water-level recovery is measured. Okada (1985) made this measurement at one station, Fukaura, and recovered the original tsunami height from the tide-gauge records. Satake *et al.* (1988) made systematic measurements at 40 tide-gauge stations in northeast Japan. Their observed responses can be used to correct tide-gauge records to recover the original tsunami waveforms.

Figure 4 shows the example of the recovery process of tsunamis waveforms. The left column is the tsunami record at Yoshioka from the 1983 Japan Sea earthquake and the right is at Hanasaki from the 1968 Tokachi-oki earthquake. The top row shows the original tide-gauge records. The tidal components are easily estimated from the longer portion of the tide-gauge records and are removed first (middle row). The bottom waveforms are corrected for the tide-gauge responses. The correction changed the waveforms significantly for the record at Yoshioka, but does not affect so much the record at Hanasaki where the tsunami period is longer and the tide-gauge response is better.

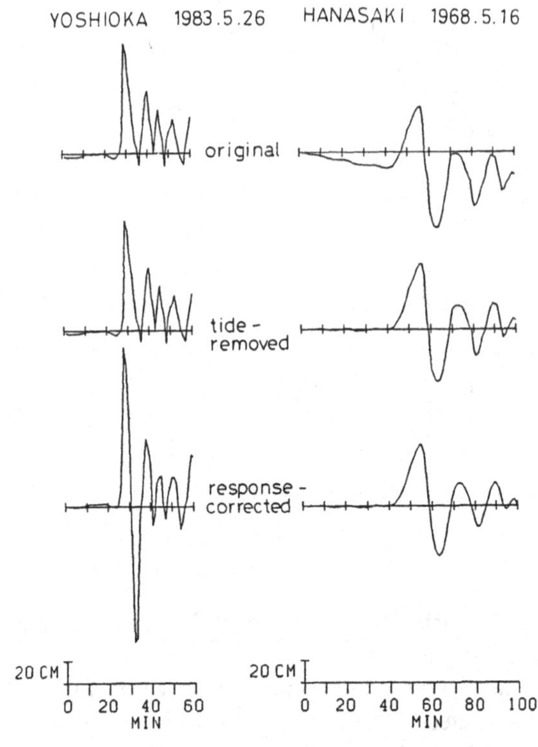

Fig. 4. Tsunami waveforms from the 1968 Tokachi-oki and the 1983 Japan Sea tsunamis (Satake, 1989). Top ones are the original tide-gauge records, the middle ones are the tide-removed data and the bottom are the waveforms corrected for the tide-gauge responses.

4. Inversion of Tsunami Waveforms

We have shown in the previous two sections that both the propagation effects and the instrumental responses can be removed from tide-gauge records. The remaining unknown is the source. Recent developments in seismology have shown that fault motions of large earthquakes are not uniform but usually have some patches where slip is large (e.g., Kanamori, 1986; Kikuchi and Fukao, 1987). Consequently, the crustal deformation on the ocean bottom, i.e., the initial wave field for tsunami propagation, must also be heterogeneous. We use tsunami waveforms to estimate the spatial slip distribution on the fault plane in the following way (Figure 5).

The fault plane is divided into small subfaults and the deformation on the ocean bottom is computed for each subfault with a unit amount of slip. Using this displacement field as the initial condition, tsunami waveforms are computed by a finite-difference method using actual bathymetry. The observed tsunami waveforms are expressed as a superposition of the computed waveforms as follows,

$$A_{ij}(t)x_j = b_i(t),$$

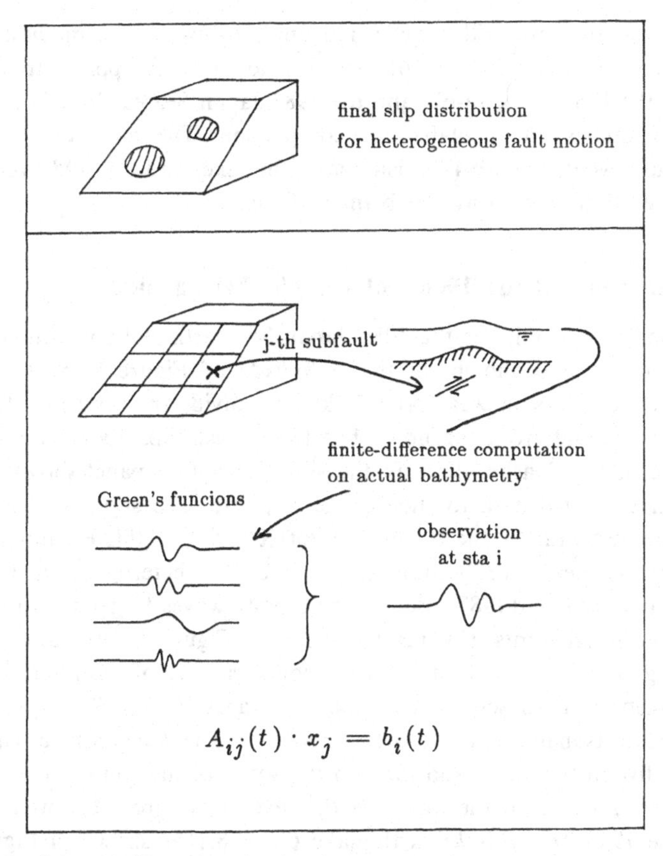

Fig. 5. Schematic illustration showing the inversion of tsunami waveforms to estimate heterogeneous fault motion.

where A_{ij} is the computed waveforms, or Green's functions, at the ith station from the jth subfault, x_j is the amount of slip on the jth subfault, and b_i is the observed tsunami waveform at the ith station. The slip x_j on each subfault can be estimated by a least-squares inversion of the above set of equations.

We first conducted numerical experiments to estimate the spatial resolution of the inversion method (Satake, 1989). The experiment was made in the Japan Sea where the 1983 Japan Sea earthquake ($M = 7.7$) tsunami occurred. For this earthquake, forward modelling using seismic waves and tsunamis was made by Satake (1985), so the computed waveforms from his final model were used as observations to simulate the inversion. This experiment shows that the spatial resolution of source complexity is about 30 km, when the grid spacing of finite-difference computation is 2.5 km and the time interval of digitized tide-gauge record is 1 min. Since the fault area of earthquakes with $M = 7.5$ is typically 100 km or larger, the present method will bring information on the heterogeneous fault motion of such large earthquakes.

In the following sections, we will show some applications of the method to several earthquakes near Japan (Figure 6). The method is first applied to two recent earthquakes, the 1968 Tokachi-oki and the 1983 Japan Sea earthquakes, for which seismological studies are available for comparison. The method is then applied to an unusual event, the 1984 Torishima earthquake, and two old events, the 1944 and 1946 earthquakes along the Nankai Trough.

5. The 1983 Japan Sea and the 1968 Tokachi-oki Earthquakes

The slip distribution on the fault for the 1983 Japan Sea earthquake is estimated from tsunami waveforms and shown on the right side of Figure 7. Since the spatial resolution of the inversion was about 30 km, we divide the fault into four subfaults. The third subfault from the north has the largest slip, about twice as large as that on the others. Dashed lines in the slip-distribution panel shows the estimate from the uncorrected data for tide-gauge response. The slip distribution would be underestimated without the tide-gauge correction. For this earthquake, several seismological studies have been made to infer the heterogeneous fault motion. Among them, Houston (1987) used far-field body waves to get spatio and temporal distribution of subevents. Her results, shown in Figure 7 (left side), also showed that the largest subevent is located in the southern part of the fault. This large slip is confirmed by both seismic and tsunami waves. Figure 8 shows the observed and computed tsunami waveforms for this event. The waveforms at eight tide-gauge stations, five in the north and three in the south of the source area, are used in the inversion. As seen in the figure, both waveforms agree very well.

The result for the 1968 Tokachi-oki earthquake ($M_s = 8.1$) is shown in Figure 9 as well as the results of the seismic wave analysis. The arrows show the epicenter, a reference point, for three pictures. Two common features can be seen in these results; the maximum slip is located northwest of the epicenter and the slip south of the epicenter is very small or zero. Again, the result from tsunami inversion is similar to those from seismic wave analyses. Since the tsunami represents longer-period motion than seismic waves, the results from tsunami are not necessarily the same as those from seismic waves. The similar results from tsunami and seismic waves in the above two events indicate that the frequency characteristics of the source motion is similar for the tsunami and seismic wave period ranges.

6. Unusual Earthquake

The Torishima earthquake of 13 June 1984 caused unusually large tsunamis for its size. This event was located on a shallow ridge system on the back-arc side of the Japan Trench (Figure 10). The seismic magnitudes (both surface wave magnitude M_s and moment magnitude M_W) were 5.6. At Yaene on Hacijo-jima, 190 km north of the epicenter, the visually observed tsunami height was 130 to 150 cm,

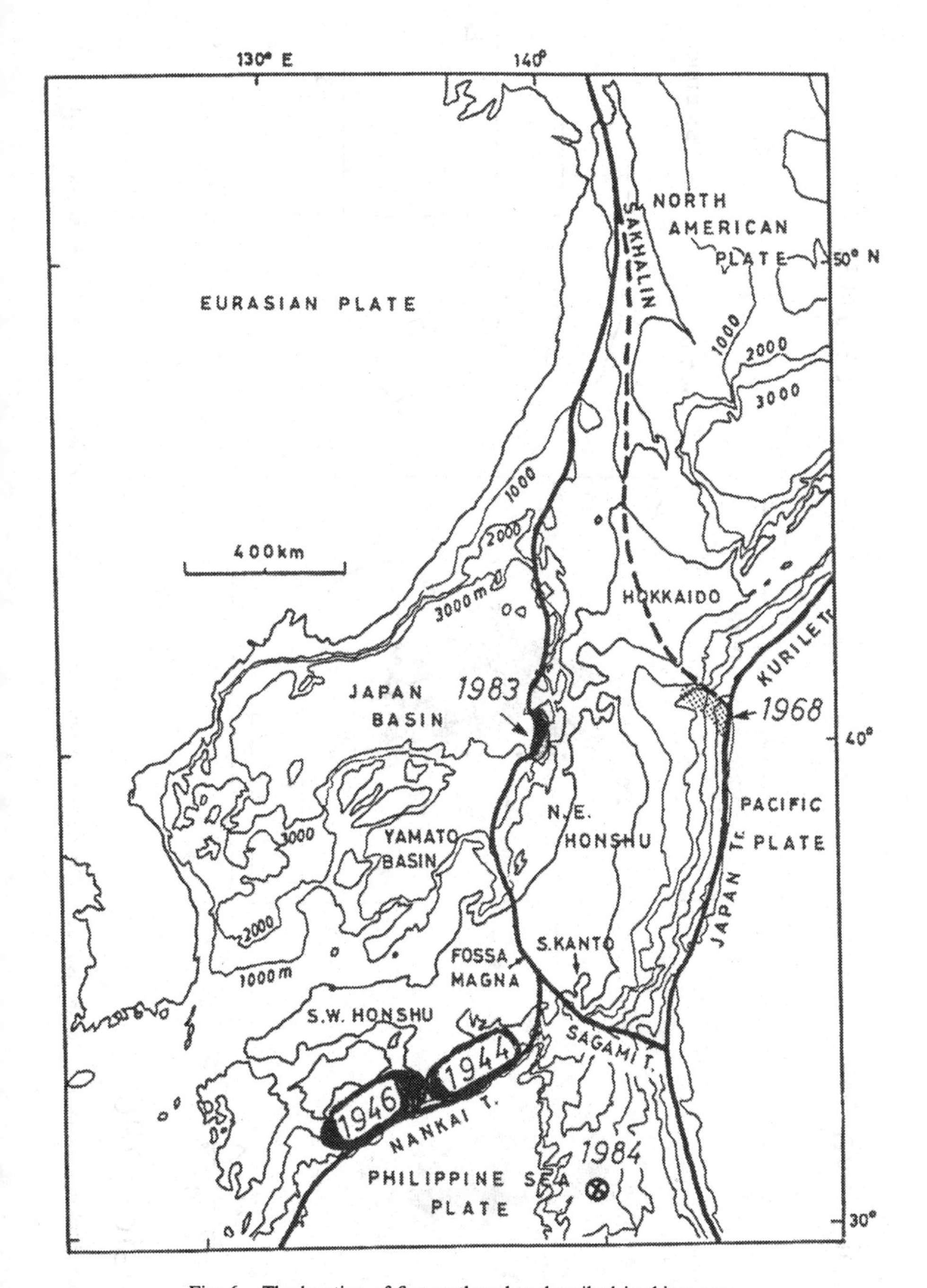

Fig. 6. The location of five earthquakes described in this paper.

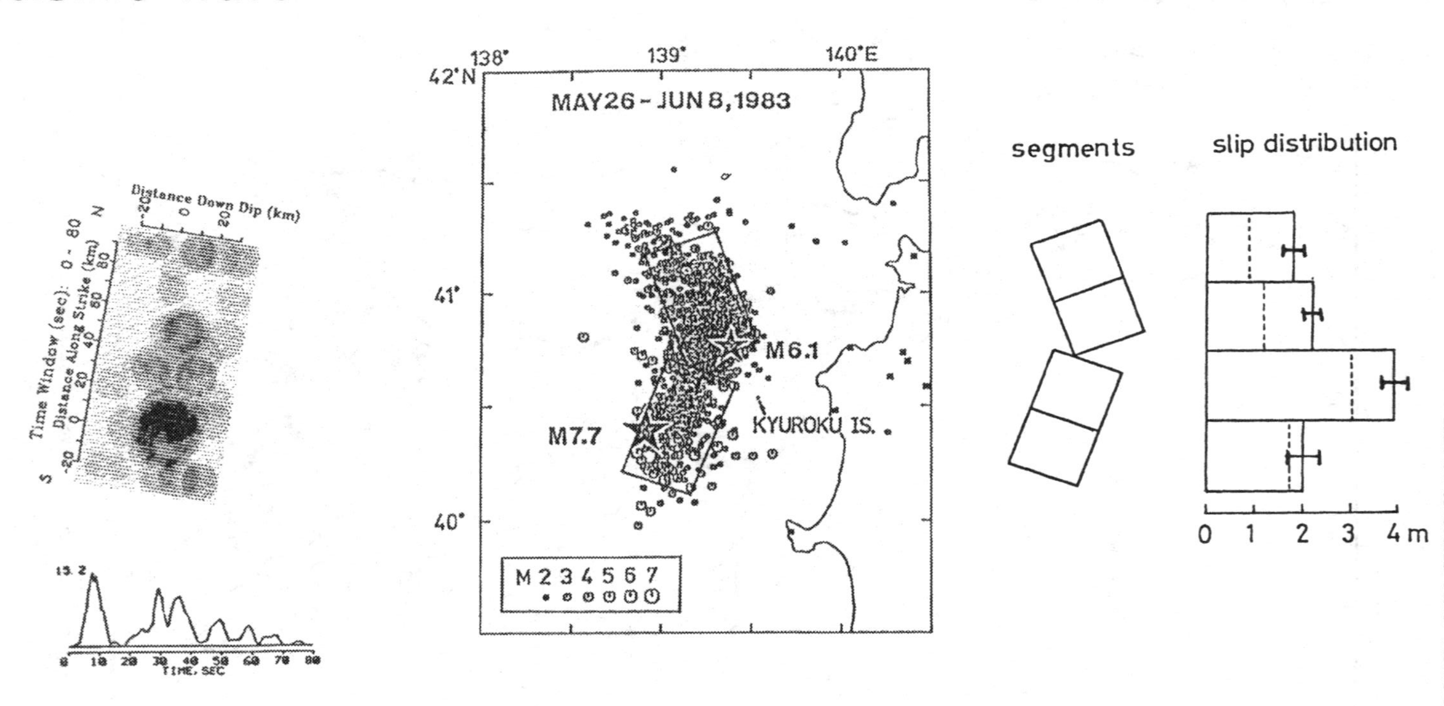

Fig. 7. Slip distribution on the fault for the 1983 Japan Sea earthquake estimated from seismic waves (Houston, 1987) and by tsunami waves (Satake, 1989).

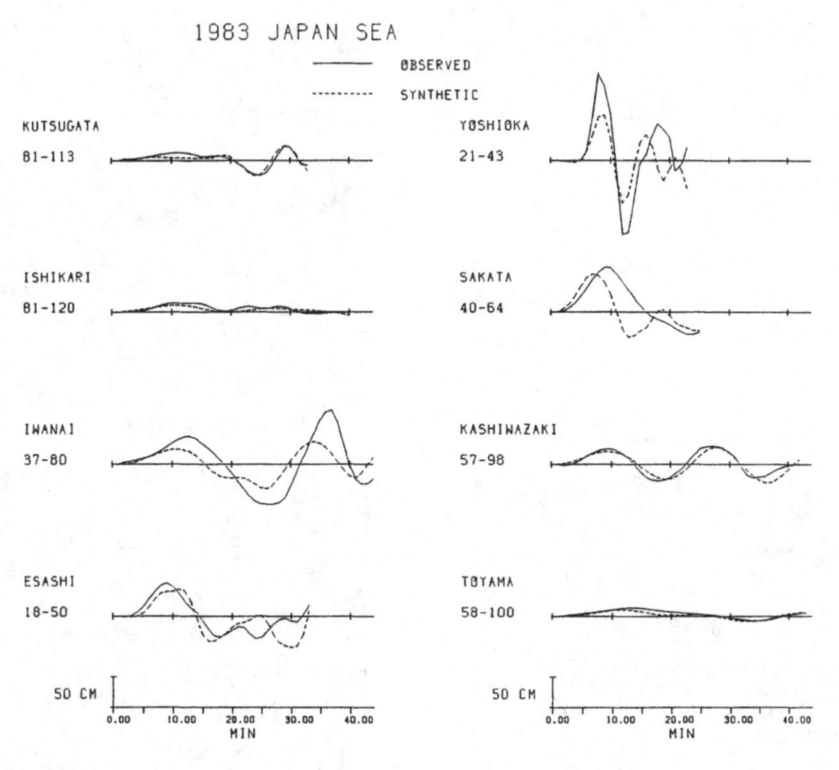

Fig. 8. Comparison of the observed (solid curves) and computed (dashed curves) waveforms for the 1983 Japan Sea earthquake tsunami (Satake, 1989).

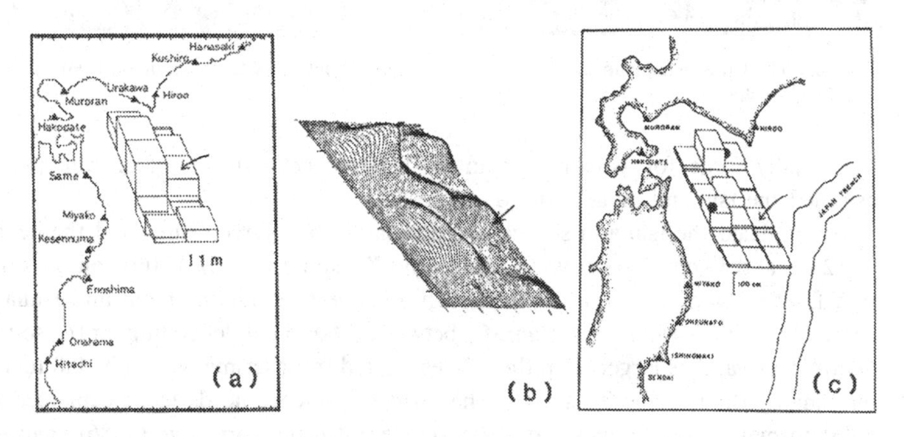

Fig. 9. The slip distributions for the 1968 Tokachi-oki earthquake. (a) is from tsunami waves by Satake (1989), (b) is from seismic body waves by Kikuchi and Fukao (1985), and (c) is from seismic surface waves by Mori and Shimazaki (1985).

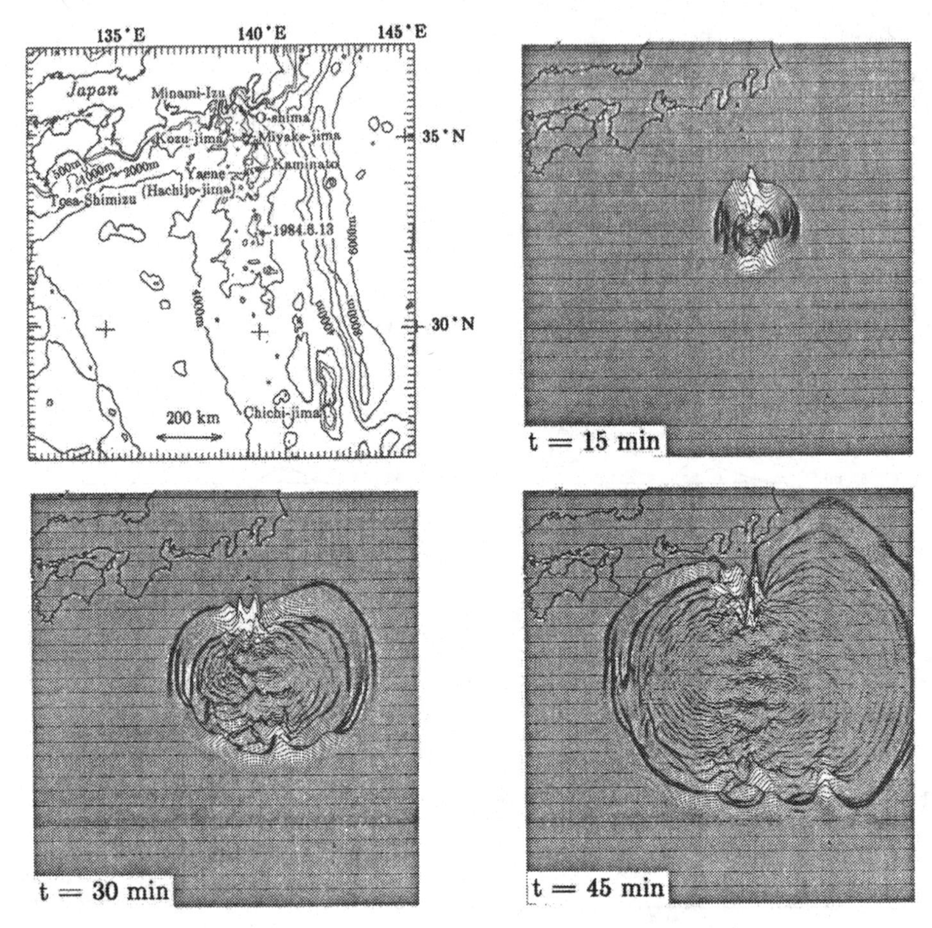

Fig. 10. The bathymetry map and snapshots at three different times of the tsunami from the 1984 Torishima earthquake.

abnormally large for tsunamis from an earthquake with $M = 5.6$. Abe (1989) assigned the tsunami magnitude M_t to be 7.3.

We modeled the tsunami source and estimated the source radius and the height as 12 km and 13 cm, respectively (Satake and Kanamori, 1990). If this source is due to a fault motion, an empirical relationship between seismic moment and tsunami energy by Kajiura (1981) gives an M_W between 6.4 and 6.9 depending on the source depth. This value is larger than the M_W estimated from seismic waves by 0.8 to 1.3, but smaller than M_t by 0.4 to 0.9. The former difference needs to be explained by an abnormal source probably at a shallow depth. Kanamori *et al.* (1986) analyzed the seismic surface waves and found that this event can be best explained by a sudden injection of water-magma mixture into shallow sediments.

The latter discrepancy in magnitudes must be due to the propagation effects, because the tsunami magnitude M_t does not consider the propagation effects while we do in our numerical computation. Figure 10 shows the snap-shots of the tsunami wave-field at 15, 30 and 45 min after the origin time. Up to 15 min, the tsunami propagates at an almost uniform velocity and the amplitude is almost uniform at all the directions. At 30 min, the propagation in the northeast direction, where the deep Japan Trench lies and the tsunami velocity is high, is more progressed than in other directions. The amplitude is larger in the northern direction where the water is shallow on the ridge system. This feature is clearer at 45 min; the focusing of energy on the shallow ridge and defocusing in the deep trench are visible. Abnormally large tsunamis were observed in the ridge direction, which resulted in the large tsunami magnitude M_t. Thus, the abnormality of the Torishima earthquake is a result of both source and path effects.

7. Old Earthquakes

Tsunami waveform data are most useful for the study of old earthquakes. Examples shown here are the earthquakes along the Nankai Trough, Southwest Japan (Figure 11). The Nankai Trough is a typical subduction zone where the Philippine Sea Plate is subducting beneath the Eurasian Plate. The most recent large earthquakes along the Nankai Trough are the 1944 Tonankai and the 1946 Nankaido earthquakes (both $M = 8.2$). Previous to these are the two 1854 Ansei earthquakes

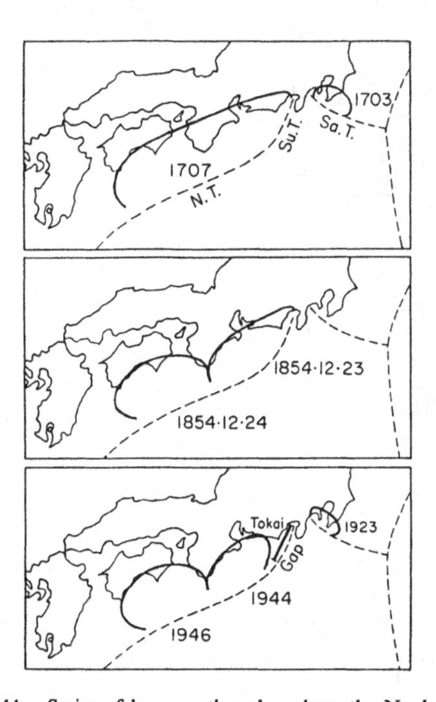

Fig. 11. Series of large earthquakes along the Nankai Trough, Japan (Mogi, 1981). Top is the 1707 Hoei earthquake, middle is the 1854 Ansei earthquakes, and bottom is the 1944 Tonankai and 1946 Nankaido earthquakes.

(both $M = 8.4$) and 1707 Hoei ($M = 8.4$) earthquake. Thus earthquakes or pairs of them with a magnitude larger than 8 have repeated roughly about every 100 years. However, the rupture pattern seems different in each cycle; a single earthquake ruptured the whole region in 1707, but two events occurred in the later sequences. Even for the most recent sequence in the 1940s, no seismogram good enough to examine the source process is available. On the other hand, many tsunami waveform data with quality comparable to modern records are still available from the 1944 and 1946 events.

We estimated the slip distribution of the 1944 Tonankai and 1946 Nankaido earthquakes by the waveform inversion. The obtained slip distributions are not uniform as shown in Figure 12. The largest slip is located at the deepest parts of the fault plane. The shallowest part of the fault, under the accretionary wedge, also slipped at the time of the mainshock. This result provides evidence that the slip beneath the accretionary wedge is coseismic and can be used to estimate the tsunami potential at geologically similar subduction zones such as the Cascadia subduction zone. This feature of heterogeneous fault motion beneath the accretionary wedge is first clarified by the present tsunami analyses.

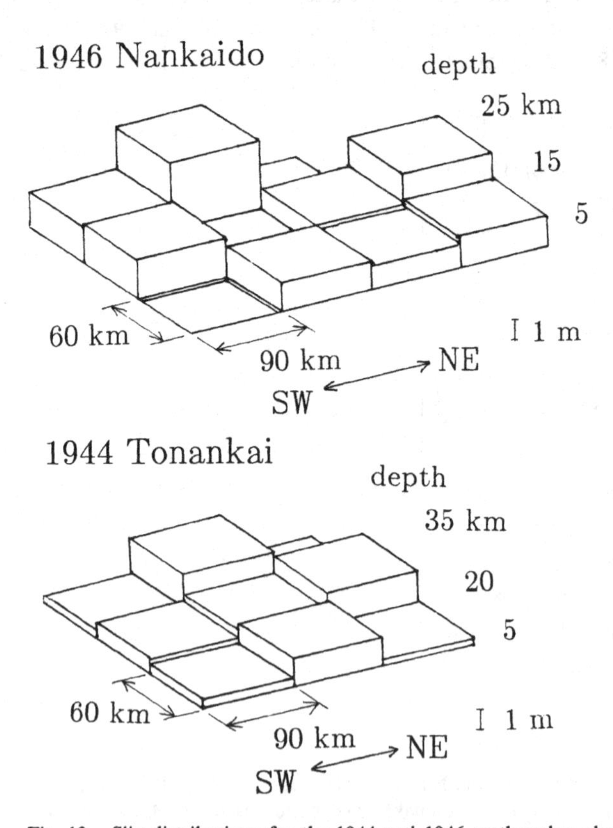

Fig. 12. Slip distributions for the 1944 and 1946 earthquakes along the Nankai Trough.

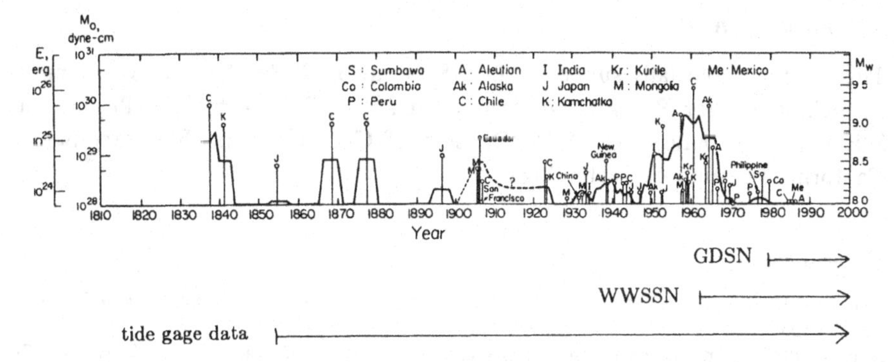

Fig. 13. Temporal variation of energy release in earthquakes and the duration of available data.

8. Discussion

We have shown that tsunami waveforms provide independent and reliable information on earthquake source processes. We also demonstrated that tsunami waveforms are most useful for the study of unusual earthquakes or old events. Figure 13 shows the temporal change of the energy release from earthquakes since the 19th century. As seen in the figure, many large or great earthquakes occurred before 1970. Examination of these is very important to understand the nature of such large earthquakes as well as the temporal variation of global seismicity, because the recurrence intervals of such earthquakes are 100 years or longer. At the bottom of the figure are shown the available periods for digital and analog global seismic network data and tide-gauge records. Although a high dynamic-range, global digital seismic network is in operation today, they provide seismograms from events only after 1980. World-wide Standard Seismic Network (WWSSN) has been in operation since 1960 and is used extensively for seismic research. Although it recorded most of the great earthquakes in the 1960s, the period of its operation is only a quarter of century, much shorter than the typical recurrence times. Tide-gauge data, on the other hand, are available as far back as 1850s for trans-Pacific tsunamis. For example, the tsunamis from the 1854 Ansei earthquakes in the Nankai Trough are recorded at the three stations on the west coast of North America. If the present waveform inversion method is extended to such trans-Pacific tsunamis, it will help to solve many important seismological problems such as temporal changes in long-term global seismicity or variable rupture mode of great earthquakes.

We made preliminary computations of the trans-Pacific tsunami from the 1964 Alaskan earthquake to see the resolvability of heterogeneous fault motion. Two previously proposed fault models are tested. The computed tsunami waveforms from these fault models at far stations such as in Japan or California are notably different, indicating that some details of the fault motion are resolvable from the observed tsunamis across the Pacific ocean.

Acknowledgements

This work was partially supported by NSF grant EAR-89-15987. K.S. is supported by Fellowship for Research Abroad from Japan Society for the Promotion of Science. Contribution No. 4815, Division of Geological and Planetary Sciences, California Institute of Technology.

References

Abe, K.: 1979, Size of great earthquakes of 1837-1974 inferred from tsunami data, *J. Geophys. Res.* **84**, 1561–1568.

Abe, K.: 1989, Quantification of tsunamigenic earthquakes by the M_t scale, *Tectonophysics* **166**, 27–34.

Comer, R. P.: 1984, Tsunami generation: a comparison of traditional and normal mode approaches, *Geophys. J. R. Astr. Soc.* **77**, 29–41.

Hatori, T.: 1969, Dimensions and geographical distribution of tsunami sources near Japan, *Bull. Earthq. Res. Inst.* **47**, 185–214.

Houston, H.: 1987, Source characteristics of large earthquakes at short periods, Ph.D. thesis, California Institute of Technology, Pasadena, 129 pp.

Kajiura, K.: 1981, Tsunami energy in relation to parameters of the earthquake fault model, *Bull. Earthq. Res. Inst.* **48**, 835–869.

Kanamori, H.: 1986, Rupture process of subduction-zone earthquakes, *Ann. Rev. Earth Planet. Sci.* **14**, 293–322.

Kanamori, H., Ekstrom, G., Dziewonski, A., and Barker, J. S.: 1986, An anomalous seismic event near Tori Shima, Japan – A possible magma injection event, *Eos Tran. Amer. Geophys. Union* **67**, 1117.

Kikuchi, M. and Fukao, Y.: 1985, Iterative deconvolution of complex body waves from great earthquakes – The Tokachi-oki earthquake of 1968, *Phys. Earth Planet. Inter.* **37**, 235–248.

Kikuchi, M. and Fukao, Y.: 1987, Inversion of long-period P-waves from great earthquakes along subduction zones, *Tectonophysics* **144**, 231–247.

Mogi, K.: 1981, Seismicity in western Japan and long-term earthquake forecasting, in D. W. Simpson and P. G. Richards (eds.), *Earthquake Predictions, An International Review*, American Geophysical Union, Washington D.C., pp. 43–51.

Mori, J. and Shimazaki, K.: 1985, Inversion of intermediate-period Rayleigh waves for source characteristics of the 1968 Tokachi-oki earthquake, *J. Geophys. Res.* **90**, 11374–11382.

Murty, T. S.: 1977, Seismic sea waves – Tsunamis, *Bull. Fish. Res. Board Canada*, **198**.

Okada, M.: 1985, Response of some tide-wells in Japan to tsunamis, *Proc. Internat. Tsunami Sympos.*, Institute of Ocean Sciences, Sidney, Canada, pp. 208–213.

Okel, E. A.: 1988, Seismic parameters controlling far-field tsunami amplitudes: a review, *Natural Hazards* **1**, 67–96.

Satake, K.: 1985, The mechanism of the 1983 Japan Sea earthquake as inferred from long-period surface waves and tsunamis, *Phys. Earth Planet. Inter.* **37**, 249–260.

Satake, K.: 1988, Effects of bathymetry on tsunami propagation: application of ray tracing to tsunamis, *Pageoph* **126**, 27–36.

Satake, K.: 1989, Inversion of tsunami waveforms for the estimation of heterogeneous fault motion of large submarine earthquakes: 1968 Tokachi-oki and 1983 Japan Sea earthquakes, *J. Geophys. Res.* **94**, 5627–5636.

Satake, K. and Kanamori, H.: 1990, Quantification of abnormal tsunamis caused by June 13, 1984 Torishima, Japan, earthquake. Submitted to *J. Geophys. Res.*

Satake, K., Okada, M., and Abe, K.: 1988, Tide gauge response to tsunamis: Measurements at 40 tide gauge stations in Japan, *J. Marine Res.* **46**, 557–571.

Natural Hazards 4: 209–220, 1991.

Tsunami Bore Runup

HARRY H. YEH
Department of Civil Engineering, FX-10, University of Washington, Seattle, WA 98195, U.S.A.

(Received: 28 December 1989; revised: 21 May 1990)

Abstract. Nearshore behaviors of tsunamis, specifically those formed as a single uniform bore, are investigated experimentally in a laboratory environment. The transition process from tsunami bore to runup is described by the 'momentum exchange' process between the bore and the small wedge-shaped water body along the shore: the bore front itself does not reach the shoreline directly, but the large bore mass pushes the small, initially quiescent water in front of it. The fluid motions near the runup water line appear to be complex. The complex flow pattern must be caused by irregularities involved in the driving bore and turbulence advected into the runup flow. Those experimental results suggest that the tsunami actions at the shoreline involve significant mean kinetic energy together with violent turbulence. Even though the behaviors of bore motion were found to be different from those predicted by the shallow-water wave theory, the maximum runup height appears to be predictable by the theory if the value of the initial runup velocity is modified (reduced). Besides the friction effect, this reduction of the initial runup velocity must be related to the transition process as well as the highly interacting three-dimensional runup motion.

Key words. Bore, turbulence, wave runup, tsunami, long wave, coastal zone.

1. Introduction

As tsunamis approach a shore, they sometimes break offshore and form bores near the shoreline. Such bore formations of tsunamis were evidently observed at Hilo, Hawaii, in the 1960 Chilian Tsunami, and, recently, along the western coast of Akita Prefecture in the 1983 Nihonkai-Chubu earthquake tsunami. Since offshore wave breakings induce energy dissipation of incoming waves, the formation of bores may result in smaller runup height and, consequently, causes less damage than the case of nonbreaking tsunami attack with the equivalent energy magnitude. Contrary to this speculation, the events associated with tsunami bores were recorded to cause substantial damage in the coastal areas. For example, the tsunami caused 61 deaths in the city of Hilo, Hawaii, whereas 100 deaths were reported from the 1983 Nihonkai-Chubu earthquake tsunamis in Japan. In Hilo, rocks weighing more than 20 tons were removed from a sea wall and carried as far as 200 m inland (Cox and Mink, 1963). Similarly, as shown in Figure 1, the 1983 Nihonkai-Chubu earthquake tsunamis caused the destruction of protective concrete units weighing 4 tons and scattered them widely with the maximum inshore transport of 135 m. It is evident that substantial tsunami energy is released at the shore in both events.

Most of the mathematical models for tsunami propagation and runup are based on the shallow-water wave theory, i.e. the depth-integrated equations of momentum and mass conservation with the assumption of a hydrostatic pressure field. (The

Fig. 1. Protective concrete units weighing 4 tons were scattered by the 1982 Nihonkai-Chubu earthquake tsunami in Japan.

shallow-water wave theory is thought to give a good approximation of tsunami motions, since tsunamis have a horizontal characteristic length scale much larger than the vertical length scale, so that the pressure field is considered to be close to hydrostatic.) As far as a simple nonbreaking wave is concerned (e.g. solitary and cnoidal waves), the theory can predict not only the maximum runup height but also the runup process itself on a plane beach (Synolakis, 1991), hence the basic physics of the tsunami runup can be predicted based on the present stage of the (shallow-water) model. On the other hand, once the incoming tsunami breaks offshore, the shallow-water wave theory poorly predicts the runup motions; the maximum runup heights observed in a laboratory (e.g. Miller, 1968) are always less than the prediction based on the shallow-water wave theory. Because the discrepancy appears to increase as the beach slope decreases, Hibberd and Peregrine (1979) attributed this to the friction effects. Note that the smaller the slope the longer the runup length, whereas the theoretical (frictionless) prediction of the maximum runup height is independent of the beach slope. Later, Packwood and Peregrine (1981) modified Hibberd and Peregrine's (1979) numerical model by incorporating the friction effects with the Chézy friction term. While some improvements in agreement were achieved, their results indicate that the friction effects alone cannot be the explanation for the discrepancy in the maximum runup height; the numerical predictions with a reasonable value of the Chézy friction coefficient still considerably exceed Miller's (1968) experimental results, especially in the cases of mild beach slopes. Hence, contrary to an unbroken tsunami, the basic physics of a tsunami-bore runup is far from being completely understood.

When an incoming tsunami breaks offshore and becomes a bore near the shore, some of the wave energy must be dissipated offshore via turbulence during the propagation. Yet, as stated, there is much evidence that the tsunami bore can cause significant structural damages and casualties. In spite of the devastating effects, the maximum runup height of the tsunami bore is lower than the equivalent unbroken tsunami runup. This seemingly contradicting behavior of a tsunami bore and its destructive consequence at the shoreline are explored in this paper by performing experiments in a controlled laboratory environment. Discussions of similar topics from the fluid mechanics view point are found in the paper by Yeh *et al.* (1989).

2. Experiments

Experiments were performed in a 9.0 m long, 1.2 m wide, and 0.9 m deep wave tank as shown in Figure 2. A bore is generated by lifting the 12.7 mm thick aluminum plate gate which initially separates the quiescent water on the beach from the deeper water behind the gate. The gate is lifted with the aid of a pneumatic cylinder (10.2 cm bore diameter) which is electrically activated by a single solenoid valve with an operating air pressure of 650 kPa, i.e. the maximum lifting force is 5.3 kN. The system is capable of lifting 20 cm of the gate travel distance in 0.0708 ± 0.0012 sec. Hence, bores can be generated by almost instantaneous openings of the gate in a repeatable manner ($\pm 1.7\%$ error). The distance from the gate to the initial shoreline is 114 cm which is approximately the propagation distance of $20\,\eta_0$, where η_0 is a typical bore height offshore in the horizontal bottom region. On the other hand, the tank width is 120 cm, hence the side wall effects should be negligible for the experimental data obtained along the center line of the tank.

The bore generating system was tested with a variety of initial conditions. It was visually observed that fully developed bores can be generated when $h_1/h_0 > 2.0$, where h_0 and h_1 are the initial water depths in front of and behind the gate, respectively. In the case of $h_1/h_0 < 2$, the generated bores were undular (although the leading wave is breaking at its crest when $1.6 < h_1/h_0 < 2$); this is because linear effects of frequency dispersion become significant in comparison to the nonlinear

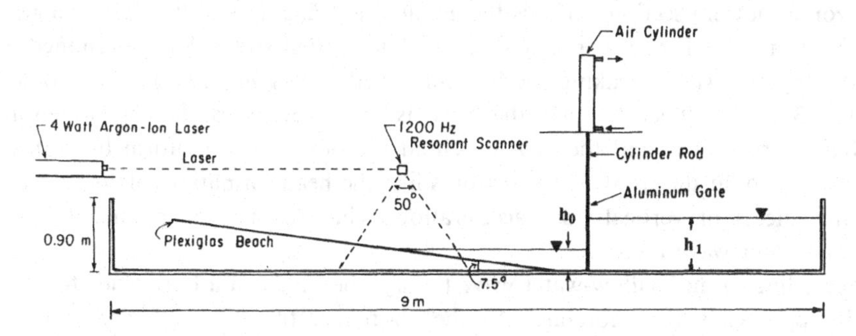

Fig. 2. A schematic view of the experimental apparatus.

effects. On the other hand, when $h_1/h_0 > 2.8$, the behavior of generated bores appeared to be too transient for the measurements. This must be due to the limited propagation distance available and the finite time involved in lifting the gate, i.e. there is insufficient time for the bore to develop before it reaches the shore. Nonetheless, for bores generated with $2.0 < h_1/h_0 < 2.6$, the propagation distance appears long enough to form a fully developed bore on the beach.

A 4 Watt Argon-ion laser is used for the visualization of a bore profile. The emitted laser beam is converted to a thin sheet of laser light through a resonant scanner. The scanner is capable of sweeping the beam at 1200 Hz with a maximum of a 50° peak-to-peak angle. The generated laser sheet is projected from above in the cross-shore (longitudinal) direction along the center line of the tank. This illuminates the vertical longitudinal plane of the water dyed with fluorescein. This type of flow visualization is often called the laser-induced fluorescent method.

Velocities of bore propagation are measured by an array of water sensors, with a sampling rate of 1250 Hz, along the center line of the tank. In the offshore region, the water sensors were placed 5 cm apart; each sensor tip was placed approximately 1 mm above the initial quiescent water level. An attempt was made to use the same method to measure the velocities of the runup tip on the beach. A difficulty arose with the formation of a very thin sheet of runup water; the flow disturbance caused by a sensor rod (1.6 mm in diameter) influences the adjacent sensor. To minimize this difficulty, the runup velocities were measured by an array of sensors embedded in the beach. The sensor tips were projected no more than 1 mm from the beach surface. The flow structures near the maximum runup were recorded from above by a 35 mm photo camera.

3. Results and Discussion

A typical water-surface profile of the laboratory-generated tsunami bore is shown in Figure 3. The bore was illuminated by the laser light sheet (approximately 1 mm thick) to visualize the profile of a virtually two-dimensional vertical longitudinal plane. The initial strength of the bore represented by the offshore Froude number is $\mathbf{F} = U_0/(gh_0)^{1/2} = 1.43$, where U_0 is the bore propagation speed offshore in the horizontal bottom section and g is the acceleration due to gravity. This strength of the bore offshore is considered typical, since the initial strength is determined and limited by the wave breaking mechanism (Svendsen *et al.*, 1978). The profile in Figure 3 evidently shows that the bore is fully developed, having an entirely turbulent front face, and the profile behind the bore is not uniform but forms a distinct 'head' at the front. Local features like the head formation might be caused by the effects of vertical fluid acceleration, which cannot be predicted by the shallow-water wave theory.

According to the shallow-water wave theory, the height of a bore tends to vanish as it approaches the shoreline. At the shoreline, the fluid velocity and bore propagation velocity approach their common finite value, U^*, whereas their

Fig. 3. A longitudinal bore profile illuminated by the laser sheet. The initial shoreline is indicated by 's'. Initial Froude number, **F** = 1.43. The toe of the bore front is at 9 cm from the shoreline.

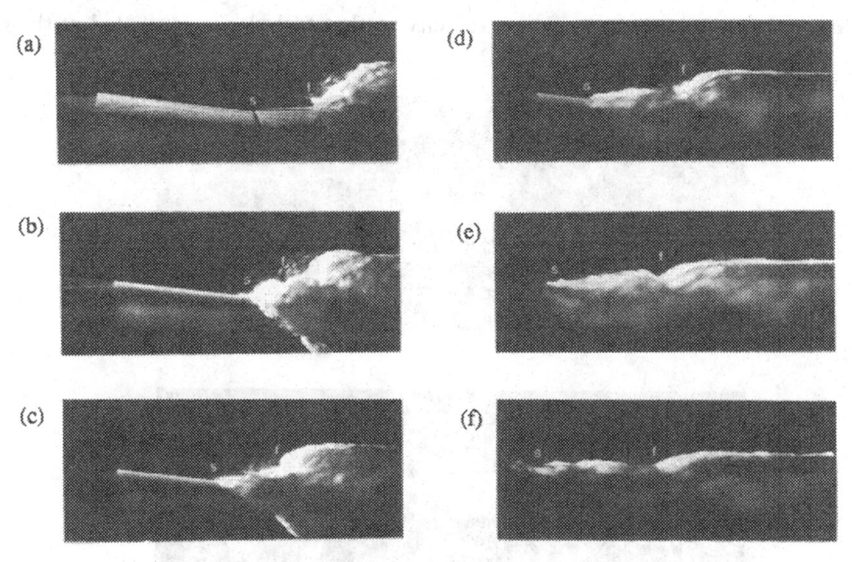

Fig. 4. Transition process from bore to runup mode. Initial Froude number, **F** = 1.43. (a) Bore approaching the shore, (b)–(e) transition, (f) runup. f, bore front; s, runup wave tip.

accelerations become infinity at the shoreline. This rapid and total conversion of potential to kinetic energy is often called 'bore collapse'. As shown in Figure 4, the prediction of 'bore collapse' did not occur in the laboratory experiments. At the shoreline, the vertical fluid acceleration becomes important in the transition process from the bore formation offshore to the tsunami runup on a beach surface, and the shallow-water wave theory cannot provide an accurate description of that process. Instead of diminishing bore height at the shoreline, the results in Figure 4 demonstrate the transition process to be like the 'momentum exchange' between the bore and the small wedge-shaped water body along the shore. The bore front itself does not reach the shoreline directly, but the bore pushes a small initially quiescent mass of water in front of it. The term 'momentum exchange' is used here to describe this transition since the process is analogous to the collision of two bodies; a fast moving large mass (i.e. bore) collides with a small stationary mass (the wedge-shaped water mass along the shoreline). Figure 4 also shows that the turbulence on the front face of the bore is advected forward onto the dry beach instead of being left behind the wave front.

Turbulence generated at the front face was observed in separate experiments. A thin layer (approximately 3 mm thick) of a slightly lighter fluid dyed with fluorescein was initially placed on the clear water. Because dominant fluid mixing is caused by advective vortices (turbulence) generated at the front, the location and behavior of turbulence can be detected by following the dyed fluid illuminated by a thin sheet of laser light. As shown in Figure 5, the turbulence generated at the front is sporadic and confined near the surface. In the offshore region, the fluid velocity is slower than the bore propagation velocity so that turbulence generated at the front

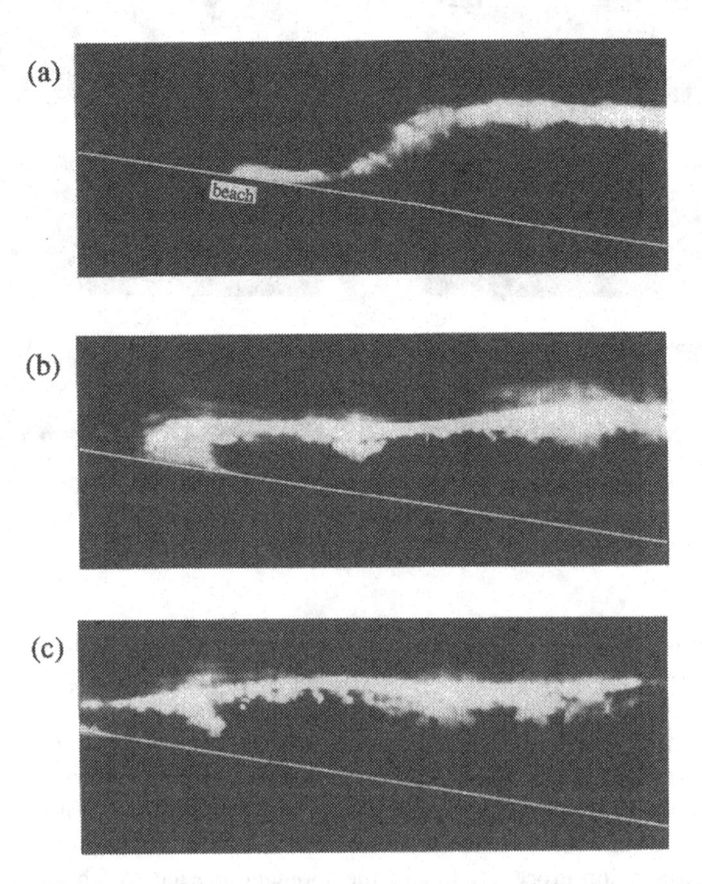

Fig. 5. Visualization of bore turbulence by the laser-induced fluorescent method. Initial Froude number, F = 1.43. The time elapsed from the generation of the bore: (a) $t = 0.76$ sec., (b) $t = 1.00$ sec, and (c) $t = 1.23$ sec.

is advected behind the front. As the bore approaches the shore, the fluid velocity approaches the propagation velocity. Since turbulence moves with the fluid velocity, the generated turbulence also moves with the velocity close to the bore propagation. Consequently, near the shoreline, the turbulence cannot be advected behind but accumulates at the front. The results shown in both Figures 4 and 5 appear to support the explanation that the accumulated turbulence at the front is released onto the dry beach during the momentum exchange process. It is emphasized that the experimental results presented here are for the event of a single uniform tsunami bore propagating into a quiescent water on an impervious plane beach. This assumed condition is relevant to a sudden large tsunami attack, but may differ from the normal wind-generated wave conditions nearshore where the effects of backwash flows created by the successive waves become important in determining the turbulent flow field.

As shown in Figure 6, the motion in the neighborhood of the runup line near its maximum was found to be complex. The complex flow pattern must be related to irregularities involved in the driving bore; i.e. turbulent flows on the bore front are irregular and three-dimensional, and the turbulence advected into the runup flow also contributes to the formation of a rough runup water surface. A single incident bore generates two successive runup motions as indicated in Figure 6. This double structure of the runup motion must be related to the transition process at the shoreline. As shown in Figure 4, two runup water masses appear to be formed; the runup of the pushed-up water mass at the shore is followed by the original bore 'front' motion which was once decelerated during the transition process.

The bore propagation velocity and the velocity of the ensuing runup water front were measured by an array of water sensors as described in Section 2. The experimental data were taken by repeating the experiments with traversing a set of

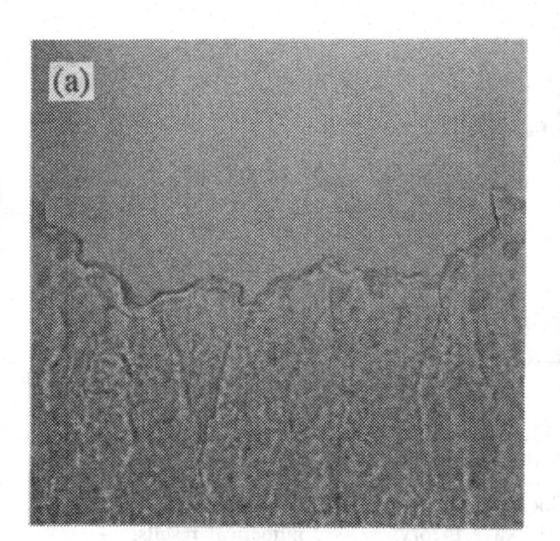

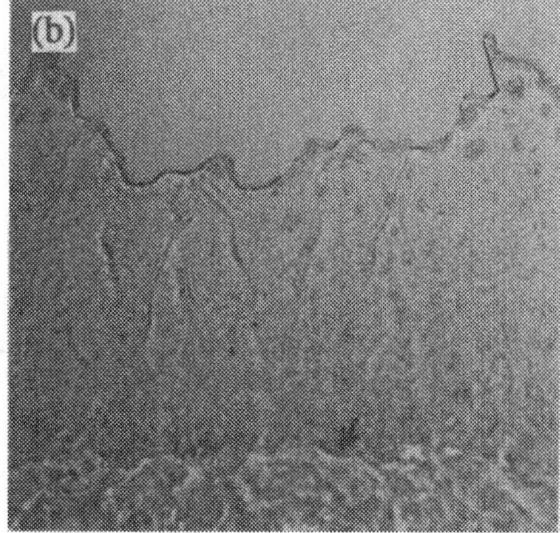

Fig. 6. Run-up waterline variation near its maximum. Time difference between (a) and (b) is 0.25 sec. Initial Froude number, $\mathbf{F} = 1.43$. The arrow in (b) identifies the successive runup motion.

eight water sensors that covers 35 cm of propagation distance. The results are presented in Figure 7 together with the analytical prediction based on the shallow-water wave theory (Ho and Meyer, 1962; Shen and Meyer, 1963) and the numerical result. The numerical result is based on the Lax-Wendroff numerical scheme and gives good agreement with the analytical prediction. Note that this agreement is not surprising, since both solutions are based on the same governing equations, i.e. the shallow-water wave equations. In spite of the repeatable bore-generation system described in Section 2, the bore propagation velocities are largely scattered in the offshore region. This scatter is not considered to be measurement errors nor a repeatability problem, but is due to the irregularities associated with the bore propagation itself. According to the observations of bore propagating on a horizontal bottom boundary (Yeh and Mok, 1990), a bore front is irregular in the transverse direction and the propagation of the front widely fluctuates. Figure 7 also indicates that the bore velocity offshore, U, decelerates faster than the prediction. Besides the frictional effects and the effects of frequency dispersion, the reason for this discrepancy is not clear. Nonetheless, this compression of the bore front as it propagates toward the shore is consistent with the formation of the bore 'head' discussed in Figure 3.

The measured results in Figure 7 show a sudden acceleration of the propagation at the shoreline, which seems to qualitatively support the occurrence of bore collapse. However, as discussed, the actual transition involves the momentum

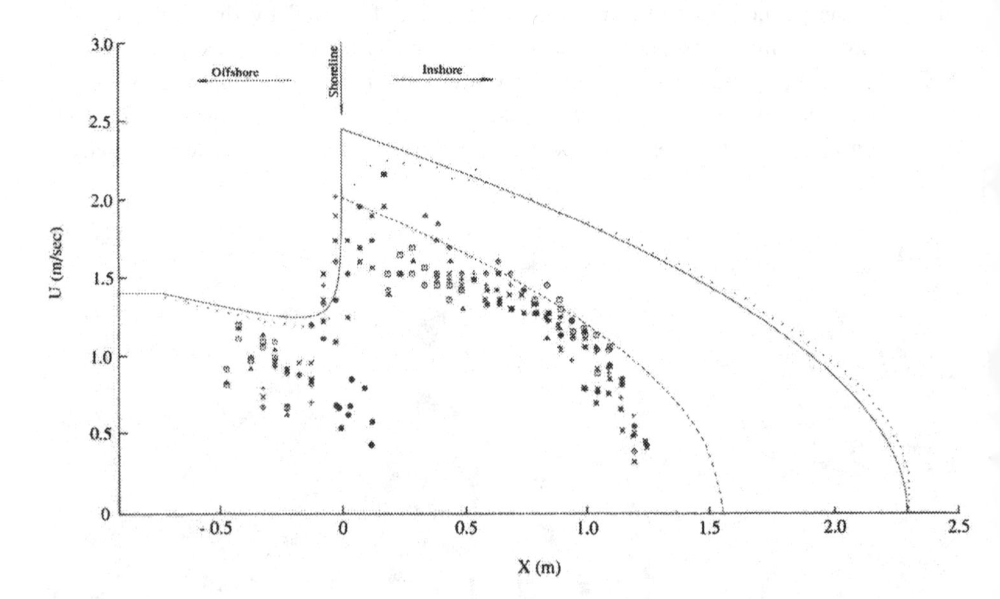

Fig. 7. Variation of bore and runup front velocities, U. Initial Froude number, $\mathbf{F} = 1.43$; ——, theoretical velocities predicted by the shallow-water wave theory; · · · · ·, numerical results; – – –, modified theoretical velocities. Different symbols denote measured velocity data taken by the repeated experiments and ● is the velocity of the 'bore front', i.e. the motion of f shown in Figure 4.

exchange between the incident bore and the small wedge-shaped water along the shoreline, but not a genuine bore collapse. The widely scattered data near the shoreline are due to this momentum exchange process. Also plotted in Figure 7 are the velocity of the 'bore front' during the transition (see Figure 4 for the location of 'bore front'); because of the momentum exchange process, the location of the bore front lags behind the runup water front. The result indicates that the 'bore front' velocity reduces during the transition process. The separation of these two runup motions supports the appearance of double runup water masses shown in Figure 6.

During the runup process, the propagation velocity is always slower than the prediction. The prediction for runup motion is made based on the assumption that the motion is governed by the gravity force only (Shen and Meyer, 1963). The initial runup front velocity is modified by using the measured maximum velocity at the shoreline and presented by the broken line in Figure 7. The measured values seem to be in fairly good agreement with the modified prediction. However, although the data are widely scattered, a careful observation of Figure 7 indicates that the runup front decelerates slower than the prediction in the region up to approximately 0.9 m inshore from the initial shoreline. Due to the viscous effect present in a real fluid environment, the bore front should have decelerated faster than the prediction of the inviscid theory. Hence, this adverse trend suggests that gravity is not the only force dominant but the pressure force is also important during the early stage of the runup. Considering the transition process at the shoreline, the runup motion is initiated by 'pushing' the water mass. This momentum exchange process is not instantaneous but rather gradual, and the runup motion forms a thick layer of flow. (The pressure force could not be effective if the runup flow had formed a thin layer.) Hence, the driving force for the runup motion appears to be the pressure gradient existing during the transition.

The velocity variations of the bore front and of the runup front for four different initial bore conditions (the initial Froude Number $\mathbf{F} = 1.31$ to 1.48) are presented in the nondimensionalized form in Figure 8; this non-dimensionalization is made in accordance with the shallow-water wave theory (Ho and Meyer, 1962). In the figure, the scaling parameter U^* is the fluid (and also propagation) velocity at the shoreline which is predicted with the initial bore condition offshore, and β is the slope of the beach. In Figure 8, the data reasonably coincide into one pattern indicating a slower propagation speed offshore than the prediction, a large scatter near the shoreline, and slower deceleration than the prediction during the earlier stage of the runup. Hence, the characterizations we made for Figure 7 are valid for the range of bore conditions but not limited to the case with the Froude number, $\mathbf{F} = 1.43$. Furthermore, in spite of the fact that detailed bore and runup behaviors are different from the predictions, collapsing all the nondimensionalized data (based on the shallow-water wave theory) into a single pattern suggests that the maximum runup height can be predicted from the initial offshore condition by the shallow-water wave theory by modifying the value of U^*. From our series of experiments, measured runup heights appear to be approximately 60% of the predicted values,

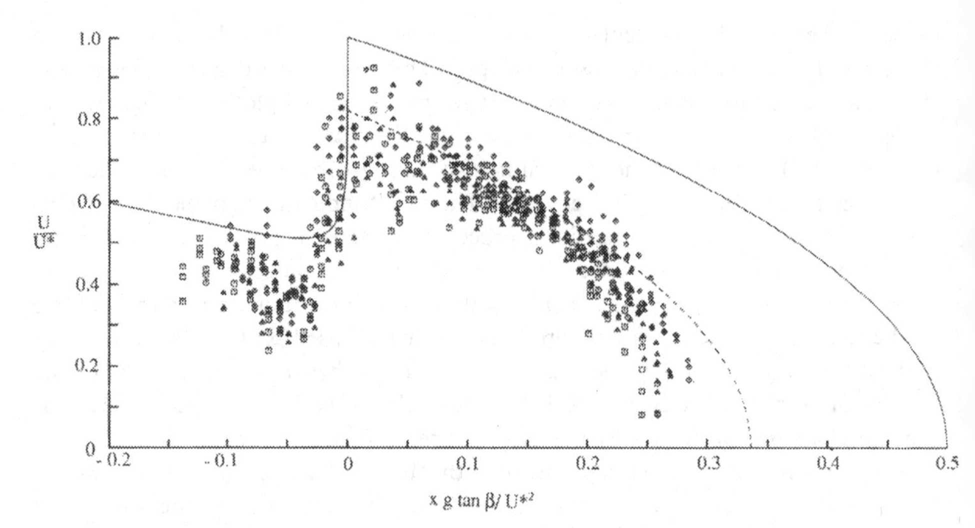

Fig. 8. Variation of bore and runup front velocities for the various initial bore strengths. □, **F** = 1,31; ○, **F** = 1.37; △, **F** = 1.43; ◇, **F** = 1.48, ——, theoretical velocities predicted by the shallow-water wave theory; – – –, theoretical velocities modified by 0.82 U^*.

which is equivalent to the reduction of U^* by 77%. Note that this specific magnitude of the reduction might be influenced by scale effects of our experimental setup and cannot be considered to be the general figure applicable for a real tsunami runup. Besides the friction effect, the reduction of the initial runup velocity should be related to the transition process; instead of sudden conversion of potential to kinetic energy ('bore collapse' as predicted by the theory), the runup water is pushed up by the gradual momentum exchange via pressure force, hence some of the incident wave energy must be reflected offshore during the transition. In addition, the three-dimensional and highly interacting complex runup motion created by irregularities involved in the driving bore could also cause additional reduction of the maximun runup height.

4. Concluding Remarks

The detailed flow transition process of a tsunami at the shoreline and the ensuing runup motion was investigated experimentally. The experiments were designed specifically for a single uniform bore propagating into an initially quiescent water on an impervious plane beach. The assumption of the flow field being quiescent initially is reasonable to a tsunami attack to a coastal region because of its greater magnitude and the longer wave length than those of the background wind-generated wave field. It is, however, emphasized that the results presented in this paper may not directly represent usual wave motions in a coastal region where a sequence of waves are continually approaching the shore.

As discussed in Section 3, violent flow actions may result at the transition of a tsunami bore to the runup. In the offshore region, turbulence generated at the front of the bore is advected behind the front because the fluid velocity is slower than the propagation velocity there. However, as the bore approaches the shore, turbulence created on the front face tends to accumulate at the front due to the convergence of the fluid and propagation velocities, i.e. at the shoreline, there is no water in front of the bore so the fluid must move with the velocity of the wave propagation. Consequently, strong turbulence is supplied to the transition process at the shoreline. Furthermore, additional turbulence is generated by the 'momentum exchange' process by pushing a small, initially quiescent water along the shoreline. This generated and accumulated turbulence is released onto the dry beach surface together with the rapidly-accelerated mean flow motion. From this view point, the transition process can cause destructive tsunami energy release at the shoreline. Hence, careful preparation, especially along the shoreline, is essential for tsunami hazard mitigation. Under-designed shoreline protection can cause adverse effects since the protection devices themselves, e.g. concrete blocks, could be washed away and become water-borne missiles as demonstrated in Figure 1. Because of its highly turbulent flow, a significant amount of sediment can be transported by a tsunami from the shoreline to the island area. On the other hand, sediment offshore is unlikely to be transported to significantly distant inland areas because, again, the fluid velocity lags behind the propagation velocity in the offshore region and no extraordinary mechanism was found for sediment transport in the offshore region.

Although the detailed behaviors of tsunami bores are different from the predictions based on the shallow-water wave theory, the maximum runup height can be predicted by the theory using a modified (reduced) value of the initial run-up velocity. Besides the friction effect, this reduction of the initial runup velocity might be related to the transition process and the highly interacting complex runup motions. (As stated in Section 1, the friction effects alone cannot be the explanation for the discrepancy in the maximum runup height.) Instead of sudden conversion of potential to kinetic energy (as predicted by the theory), the runup water is pushed up by the gradual momentum exchange via pressure force; hence, some of the incident wave energy can be reflected offshore during the transition process. The runup motion is initiated by the irregular driving bore rather than a clean two-dimensional motion. The three-dimensional runup flow motion together with the induced turbulence can dissipate the energy and reduce the maximum runup height.

Acknowledgements

The author wishes to thank Dr A. Ghazali and Mr T. McKay of the University of Washington for their assistance in the experiments. Professor N. Shuto generously provided me with the photograph shown in Figure 1. The work for this paper was supported by the U.S. National Science Foundation Grant No. CES-8715450.

References

Cox, D. C. and Mink, J. F.: 1963, The tsunami of 23 May 1960 in the Hawaiian Islands, *Bull. Seism. Soc. Am.* **53**, 1191–1209.

Hibberd, S. and Peregrine, D. H.: 1979, Surf and runup on a beach: A uniform bore, *J. Fluid Mech.* **95**, 323–345.

Ho, D. V. and Meyer, R. E.: 1962, Climb of a bore on a beach. 1: Uniform beach slope, *J. Fluid Mech.* **14**, 305–318.

Miller, R. L.: 1968, Experimental determination of runup of undular and fully developed bores, *J. Geophys. Res.* **73**, 4497–4510.

Packwood, A. R. and Peregrine, D. H.: 1981, Surf and runup on beaches: models of viscous effects, Rep. AM-81-07, Univ. of Bristol.

Shen, M. C. and Meyer, R. E.: 1963, Climb of a bore on a beach. Part 3. Run-up, *J. Fluid Mech.* **16**, 113–125.

Svendsen, I. A., Madsen, P. A. and Hansen, J. B.: 1978, Wave characteristics in the surf zone, *Proc. 16th Conf. Coastal. Eng.* 520–539.

Synolakis, C. E.: 1991, Tsunami runup on steep slopes: How good linear theory really is, *Natural Hazards* **4**, 221–234 (this issue).

Yeh, H. H., Ghazali, A., and Marton, I.: 1989, Experimental study of bore runup, *J. Fluid Mech.* **206**, 563–578.

Yeh, H. H. and Mok, K. M.: 1990, On turbulence in bores, *Phys. Fluids* **A2**, 821–828.

Natural Hazards **4**: 221–234, 1991.

Tsunami Runup on Steep Slopes: How Good Linear Theory Really Is

COSTAS EMMANUEL SYNOLAKIS
School of Engineering, University of Southern California, Los Angeles, CA 90089-2531, U.S.A.

(Received: 28 January 1990; revised: 29 June 1990)

Abstract. This is a study of the application of linear theory for the estimation of the maximum runup height of long waves on plane beaches. The linear theory is reviewed and a method is presented for calculating the maximum runup. This method involves the calculation of the maximum value of an integral, now known as the runup integral. Laboratory and numerical results are presented to support this method. The implications of the theory are used to reevaluate many existing empirical runup correlations. It is shown that linear theory predicts the maximum runup satisfactorily. This study demonstrates that it is now possible to match complex offshore wave-evolution algorithms with linear theory runup solutions for the purpose of obtaining realistic tsunami inundation estimates.

Key words. Tsunami runup, longwave runup, linear shallow water theory, swash, runup, solitary waves, cnoidal waves, Green's law.

1. Introduction

Tsunami runup is the most devastating hazard associated with tsunami waves. Yet, it has been the least understood, a manifestation of the difficulty of the underlying basic hydrodynamic problem. This involves the calculation of the shoreline motion due to a long wave defined far offshore and evolving continuously over varying depth and often breaking. The mathematical description of the problem involves calculations on moving boundaries, generally difficult to handle numerically. Wave breaking is still computationally intractable; the shoreline evolution of a breaking wave is still yet to be calculated convincingly even in the simplest cases. The uncertainty regarding the shape of a tsunami at generation complicate the problem even further.

Despite these difficulties, recent results (Guza and Thornton, 1982; Synolakis, 1986, 1987, 1988; Pelinosfsky, 1989) suggest that considerable progress can be attained using linear theory. In fact, it appears that it is now possible to calculate the maximum runup satisfactorily for practical applications and within a margin from the actual runup in the field which is comparable and often smaller than that obtained using higher-order theory and computationally intense algorithms.

In this paper, an exact analytical method is presented for calculating the maximum runup of long waves and a series of experiments is described to support the theory. Then the existing empirical runup correlations will be reinterpreted and explained using linear theory. The method uses the formalism of a long wave propagating over constant depth and encountering a sloping beach. There is now

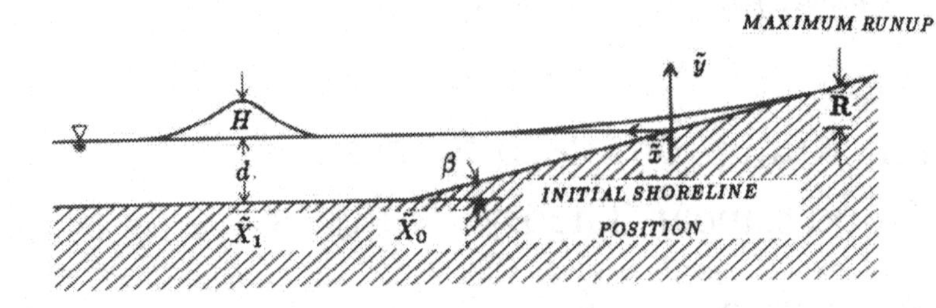

Fig. 1. A definition sketch for a tsunami runup.

consensus that this formalism describes the physical problem well. However, the approach suggested is quite general and it can be used for relatively arbitrary propagation problems, if the equations of motion are solved numerically.

The method presented is only valid for nonbreaking waves; it is now estimated (Pelinofsky *et al.*, 1989; Pelinofsky, 1989) that about 75% of the tsunamis don't break as they climb up the coastline. Another paper in this volume (Yeh, 1991) discusses the runup of tsunami bores.

2. Basic Equations and Solutions

Consider a topography consisting of a constant depth region and of a plane sloping beach of angle β, as shown in Figure 1. The origin of the coordinate system is at the initial position of the shoreline and x increases seaward. The topography $\tilde{h}_0(\tilde{x})$ is described as follows: $\tilde{h}_0(\tilde{x}) = \tilde{x} \tan\beta$ when $\tilde{x} \leqslant d \cot\beta$ and $\tilde{h}_0(\tilde{x}) = d$ when $\tilde{x} > d \cot\beta$. d is the undisturbed water depth in the constant depth region. Dimensionless variables are introduced as follows

$$\tilde{x} = xd, \quad \tilde{H} = Hd, \quad \tilde{\eta} = \eta d, \quad \tilde{h}_0 = h_0 d, \quad \tilde{u} = u\sqrt{gd},$$
$$\text{and} \quad \tilde{t} = t\sqrt{d/g}; \tag{1}$$

in general, all distances and lengths in this study are normalized by the offshore depth d.

Consider a propagation problem described by the shallow water wave equations

$$h_t + (hu)_x = 0, \tag{2a}$$

$$u_t + uu_x + \eta_x = 0, \tag{2b}$$

where $h(x) = h_0(x) + \eta(x, t)$.

2.1. *Linear Theory*

The system of Equations (2) can be linearized by retaining the first-order terms only, resulting into

$$\eta_{tt} - (\eta_x h_0)_x = 0. \tag{3}$$

When $h_0(x) = x \tan\beta$, a well-known solution is

$$\eta(x, t) = B(k, \beta) J_0(2k\sqrt{x \cot\beta})\, e^{-ikct},$$

where $B(k, \beta)$ is the amplification factor, k is the wavenumber and the celerity $c = 1$. J_0 and J_1 are the Bessel functions of the first kind of the zeroth and first-order, respectively. Keller and Keller (1964) presented another steady-state solution for the combined topography defined by Figure 1. For an incident wave of the form $\eta(x, t) = A_i\, e^{-i(x+ct)}$, they determined the amplification factor

$$B(k, \beta, A_i) = 2\, e^{-ik \cot\beta} A_i / [J_0(2k \cot\beta) - iJ_1(2k \cot\beta)].$$

Since the governing equation is linear and homogeneous, any standing-wave solutions can be used to obtain travelling wave solutions by linear superposition. (This is an important detail; the widely-held false perception that the Keller and Keller (1964) solution would only be valid for standing waves has limited its application for arbitrary waveforms.) When the incident wave is of the form $\eta(X_0, t) = \int_{-\infty}^{\infty} \Phi(k)\, e^{-ikct}\, dk$, then the wave transmitted to the beach is given by

$$\eta(X_0, t) = \int_{-\infty}^{\infty} \Phi(k) \frac{J_0(2k\sqrt{xX_0})\, e^{-ik(X_0+ct)}}{J_0(2kX_0) - iJ_1(2kX_0)}\, dk, \tag{4}$$

where $X_0 = \cot\beta$; in this normalization, this is also the distance between the initial shoreline and the toe of the beach. Setting $x = 0$ in this expression, the maximum value of the integral over all times is the maximum runup height **R** and it can be found explicitly with little computational effort. It should be noted that an equivalent equation has been presented by Pelinosfsky *et al.* (1989) who noted that "functionally, the runup height of an arbitrary pulse is described by the same formula as that of a monochromatic wave...".

Equation (4) is only valid when $x \geqslant 0$. To obtain the details of the solution when $x < 0$ one must solve the nonlinear set (2). However, Equation (4) includes the full-effects of reflection.

2.2. *Nonlinear Theory*

To solve the nonlinear set (2) for the sloping beach case, $h_0(x) = x \tan\beta$, Carrier and Greenspan (1958) introduced the following hodograph transformation

$$x = \cot\beta \left(\frac{\sigma^2}{16} - \frac{\psi_\lambda}{4} + \frac{u^2}{2}\right), \qquad t = \cot\beta \left(\frac{\psi_\sigma}{\sigma} - \frac{\lambda}{2}\right),$$

$$u = \frac{\psi_\sigma}{\sigma} \quad \text{and} \quad \eta = \frac{\psi_\lambda}{4} - \frac{u^2}{2}. \tag{5}$$

This change of variables reduces the set of Equations (2) to a single linear equation,

$$(\sigma\psi_\sigma)_\sigma = \sigma\psi_{\lambda\lambda}. \tag{6}$$

The transformation is such that in the hodograph plane, i.e. the (σ, λ) space, the shoreline is always at $\sigma = 0$. This property can be deduced easily by setting $\sigma = 0$ in (5); then $x = -\eta \cot \beta$, which is an equality only valid at the shoreline tip.

Equation (6) can be solved with standard method. Defining the Fourier transform of $\psi(\sigma, \lambda)$ as

$$\Psi(\sigma, \tilde{k}) = \int_{-\infty}^{\infty} \psi(\sigma, \lambda)\, e^{-i\lambda\tilde{k}}\, d\lambda$$

and if $\Psi(\sigma_0, \tilde{k}) = F(\tilde{k})$, then the bounded solution at $\sigma = 0$ and $\sigma = \infty$ takes the form

$$\psi(\sigma, \lambda) = \frac{1}{2\pi} \int_{-\infty}^{\infty} F(\tilde{k}) \frac{J_0(\tilde{k}\sigma)}{J_0(\tilde{k}\sigma_0)}\, e^{i\tilde{k}\lambda}\, d\tilde{k}. \tag{7a}$$

If an initial condition is available instead, one may use Hankel transform methods (see, for example, Carrier (1966)).

The process of explicitly specifying an initial or a boundary condition to Equation (6) is nontrivial. Even if initial or a boundary data are available in the (x, t) space, the process of deriving the equivalent conditions in the (σ, λ) space is very complex. These difficulties have restricted the use of the Carrier and Greenspan formalism; this is rather unfortunate, because some of the problems described can be circumvented (Carrier, 1966). Another method has been developed by Synolakis (1986, 1987) to specify a boundary condition *including reflection.* This method is briefly summarized here.

Carrier (1966) first pointed out that far from the shoreline, nonlinear effects should be rather small. The transformation equations could then be simplified by neglecting terms $\sim O(u^2)$. To the same order,

$$\psi_\lambda \ll \frac{\sigma^2}{16} \quad \text{and} \quad \frac{\psi_\sigma}{\sigma} \ll \frac{\lambda}{2}.$$

Using these approximations, the transformation equations (5) reduce to

$$u = \frac{\psi_\sigma}{\sigma}, \qquad \eta = \frac{\psi_\lambda}{4}, \qquad x = \frac{\sigma^2}{16} \cot \beta \quad \text{and} \quad t = -\frac{\lambda}{2} \cot \beta.$$

The equations are uncoupled and allow direct transition from the (σ, λ) space to the (x, t) space.

One method for specifying a boundary condition in the physical space is to use the solution of the equivalent linear problem, as given by Equation (4), at the seaward boundary, where $x = X_0 = \cot \beta$, i.e. the point $\sigma = \sigma_0 = 4$ in the (σ, λ) space. Then Equation (4) implies that $\eta(X_0, t) = \frac{1}{4}\psi_\lambda(4, \lambda)$. The boundary condition $F(\tilde{k})$ in the (σ, λ) space is determined from the linear solution (4) by repeated application of the Fourier integral theorem. Assuming that $\psi(\sigma_0, \lambda) \to 0$ as $\lambda \to \pm\infty$, then the solution of equation (6) follows

$$\psi(\sigma, \lambda) = \frac{16i}{X_0} \int_{-\infty}^{\infty} \frac{\Phi(\kappa)}{\kappa} \frac{J_0\left(\dfrac{\sigma\kappa X_0}{2}\right) e^{-i\kappa X_0(1 - \lambda/2)}}{J_0(2\kappa X_0) - iJ_1(2\kappa X_0)}\, d\kappa \tag{7b}$$

2.3. *Comparison of the Linear and of the Nonlinear Theory*

The maximum runup according to the linear theory is the maximum value attained by the wave amplitude at the initial position of the shoreline, i.e. at $x = 0$, or

$$\eta(0, t) = 2 \int_{-\infty}^{\infty} \frac{\Phi(k)\, e^{-ik(X_0 + ct)}}{J_0(2kX_0) - iJ_1(2kX_0)}\, dk. \tag{8}$$

In the nonlinear theory, the maximum runup is given by the maximum value of the amplitude at the shoreline $\eta(x_s, \lambda) = x_s \cot \beta$, where x_s is the x-coordinate of the shoreline and it corresponds to $\sigma = 0$. Notice that even though the actual wave height at the shoreline is zero, in this coordinate system everything is measured with respect to the undisturbed water depth. Using (5), one obtains that

$$\eta(x_s, t) = \frac{\psi_\lambda}{4} - \frac{u_s^2}{2} = 2 \int_{-\infty}^{\infty} \frac{\Phi(\kappa)\, e^{-i\kappa X_0(1 - \lambda/2)}}{J_0(2\kappa X_0) - iJ_1(2\kappa X_0)}\, d\kappa - \frac{u_s^2}{2}; \tag{9}$$

$u_s = dx_s/dt$ is the velocity of the shoreline tip. At the point of maximum runup u_s becomes zero. Setting $u_s = 0$ and $\sigma_0 = 0$ in the transformation equations (5) reduces these equations to

$$u = 0, \qquad \eta = \frac{\psi_\lambda}{4}, \qquad x = -\eta \cot \beta \quad \text{and} \quad t = -\frac{\lambda}{2} \cot \beta.$$

Substitution of these values in Equation (9) reduces it to Equation (8), proving that the maximum runup predicted by the linear theory is identical to the maximum runup predicted by the nonlinear theory. At the minimum rundown point, the shoreline tip also attains zero velocity, and the same argument applies again. This result was first noted by Carrier (1971) in a problem where reflection was assumed negligible. However, Carrier did not provide any comparisons for polychromatic waveforms, and this result has been largely unrecognized. As shown here, it is valid in general, even when reflection is nonzero. (Recall that linear theory assumes 100% reflection in the energy flux and it is more realistic for calculations on steep beaches than other formulations that assume no reflection.)

This runup invariance is largely unexpected, because the linear and nonlinear theory differ most at the initial position of the shoreline (*ibid.*). However, it has been known for some time in the Russian tsunami community (Pelinofsky *et al.*, 1989). Even though it is difficult to explain it in physical terms, it can also be explained by noticing that Equations (3) and (6) admit identical solutions under identical boundary conditions (Kaistrenko *et al.*, 1985); if $\eta(x, t) = F(x, t)$ is a solution of (3), then $\psi_\lambda = F(\sigma, \lambda)$ is also a solution of (6). Therefore, the maximum runup **R** – which is the maximum value of $F(0, z)$ over all z – is uniquely defined and it is the same for both equations.

The implications of this invariance would have been inconsequential in actual tsunami hazard mitigation practice, had it not been for the comparisons with laboratory data which will be described later. It will be shown that results derived with linear theory model the data satisfactorily, suggesting that dispersion is relatively unimportant, at least as far as maximum runup is concerned. This implies that the maximum runup can be calculated explicitly and directly using Equation (8).

The name 'runup integral' has been coined for the integral in Equations (8) and (9). Physically, it provides the shoreline motion. Its calculation will be discussed in the next section.

2.4. *Calculation of the Runup Integral; Examples with Cnoidal Waves*

Now two different special solutions of Equations (8) and (9) will be presented for two special classes of cnoidal waves. Cnoidal waves are exact periodic solutions of the KdV equation; a cnoidal wave propagating over constant depth is given by Svendsen (1974)

$$\eta(x, t) = y_t - 1 + H\,\mathrm{cn}^2\left(2K\left(\frac{x}{L} + \frac{t}{T}\right)\middle|\, m\right). \tag{10}$$

Here, y_t is the distance of the trough from the bottom and H, L, and T are the dimensionless wave height, wavelength, and period, respectively. $K(m)$ is the complete elliptic integral of the first kind and m is the elliptic parameter. The function $\mathrm{cn}(z \mid m)$ is the Jacobian cosine elliptic function and it can be found in Abramowitz and Stegun (1972). Cnoidal waves have two important limiting cases. As $m \to 1$, it can be shown that

$$\eta(x, t) = H\,\mathrm{sech}^2\,\gamma(x - X_1 + ct). \tag{11}$$

$c = \sqrt{1 + H}$ is the wave celerity and $\gamma = \sqrt{3H/4}$. Equation (11) is the Boussinesq profile for a solitary wave which at $t = 0$ is centered at $x = X_1$. As $m \to 0$, then

$$\eta(x, t) = \frac{H}{2} \cos 2\pi \left(\frac{x}{L} + \frac{t}{T}\right), \tag{12}$$

which is the profile of a monochromatic periodic wave.

The function $\Phi(k)$ of Equation (4) associated with solitary waves, is derived in Synolakis (1986) and it is given by $\Phi(k) = (2/3)k\,\mathrm{cosech}(\alpha k)\,\mathrm{e}^{ikX_1}$ where $\alpha = \pi/2\gamma$. Substituting $\Phi(k)$ into Equation (8) and after some nontrivial algebra (Synolakis, 1987), it can be shown that the integration result is

$$R(t) = 8H \sum_{n=1}^{\infty} \frac{(-1)^{n+1} n\, \mathrm{e}^{-2\gamma(X_1 - X_0 - ct)n}}{I_0(4\gamma X_0 n) + I_1(4\gamma X_0 n)} \tag{13}$$

(Synolakis, 1988b). The series can be simplified further by using the asymptotic form for large arguments of the modified Bessel functions. When $4X_0\gamma \gg 1$, i.e. $\sqrt{H} \gg 0.288 \tan\beta$, and, again after some asymptotics, it can be shown that its maximum value $\mathbf{R}$ is given by the

$$\mathbf{R} = 2.831\sqrt{\cot\beta}H^{5/4}. \tag{14}$$

The term *the runup law* has been coined for this equation (Synolakis, 1987). Its application is discussed later.

For other cnoidal waves, it can be shown (Synolakis, 1988a) that the runup integral reduces to

$$\eta(0,t) \approx y_t - 1 +$$

$$+ \frac{4\pi^2 H}{mK^2}\sqrt{\frac{2\pi}{L}\cot\beta}\sum_{i=0}^{\infty}\sum_{j=0}^{\infty}\frac{\sqrt{(i+j+1)}q^{i+j+1}}{(1+q^{2i+1})(1+q^{2j+1})}\times$$

$$\times\cos\left[\frac{2\pi}{L}(i+j+1)(\cot\beta - c_0 t) - \frac{\pi}{4}\right] +$$

$$+ \frac{4\pi^2 H}{MK^2}\sqrt{\frac{2\pi}{L}\cot\beta}\sum_{\substack{i=0\\ i\neq j}}^{\infty}\sum_{j=0}^{\infty}\frac{\sqrt{|i-j|}q^{i+j+1}}{(1+q^{2i+1})(1+q^{2j+1})}\times$$

$$\times\cos\left[\frac{2\pi}{L}(i-j)(\cot\beta - c_0 t) - \frac{\pi}{4}\right] +$$

$$+ \frac{2\pi^2 H}{mK^2}\sum_{i=0}^{\infty}\frac{q^{2i+1}}{(1+q^{2i+1})^2}. \tag{15}$$

As $m \to 0$, it is possible to simplify this expression. However, since it is desired to determine the behavior $\forall m$, it is necessary to evaluate the series directly and find its maximum value.

Notice that in these two examples it was possible to evaluate the runup integral (8) directly with contour integration. In practical applications, the incident wave transform $\Phi(k)$ is usually defined through the time series of the surface elevation of the incident waveform. Then it is necessary to evaluate the integral numerically; this involves a rather straightforward and standard numerical integration.

2.5. *Solitary Wave Evolution*

Classic hydrodynamic theory (Lamb, 1932) predicts that the maximum wave height $\eta_{\max} \sim h$, where h is the undisturbed local depth, when waves evolve over a sloping beach. This result is referred to as Green's law. Shuto (1973) showed that the same results can be derived as a leading order behavior from the nonlinear theory. However, none of the previous results included reflection. Here, the formalism developed in the previous sections will be used to show that Green's law is valid specifically for solitary waves, over the combined topography, and including reflection.

The maximum height at any given x is given by the maximum value of the integral in Equation (4) over all times. Substituting

$$\Phi(k) = (2/3)k \operatorname{cosech}(\alpha k)\, e^{ikX_1}$$

as before, using contour integration, and after some unpleasant but direct algebra, it is possible to show that

$$\begin{aligned}\eta(x, t) &= 2\int_{-\infty}^{\infty} \frac{k \operatorname{cosech}(\alpha k) J_0(2k\sqrt{xX_0})}{J_0(2X_0 k) - iJ_1(2X_0)k}\, dk \\ &= \frac{8\pi^2}{3\alpha^2}\sum_{n=1}^{\infty} \frac{(-1)^n n\, e^{-2\gamma\theta n} I_0(4\gamma\sqrt{xX_0 n})}{I_0(4\gamma X_0 n) + I_1(4\gamma X_0 n)}.\end{aligned} \tag{17}$$

Since it is required to find a solution for large x, i.e. away from the shoreline, we can replace the modified Bessel functions by their asymptotic expansions for large arguments. Then (17) becomes

$$\eta(x, t) = \frac{4\pi^2}{3\alpha^2}\left(\frac{X_0}{x}\right)^{1/4} \sum_{n=1}^{\infty} (-1)^n n\, e^{-2\gamma\theta' n}. \tag{18}$$

Here

$$\theta' = X_1 + X_0 - ct - 2\sqrt{xX_0}.$$

The series in (18) is seen to be a power series of the form $\Sigma(-1)^n n\mathscr{X}^n$. Since $\theta' > 0$, then the series converges; it can be verified by inspection that

$$\sum_{n=1}^{\infty} (-1)^n n\mathscr{X}^n = -\frac{\mathscr{X}}{(1+\mathscr{X})^2}.$$

The maximum of this series is $\frac{1}{4}$ and it occurs at $\mathscr{X} \approx 1$. Substituting $\pi/\alpha = \sqrt{3H}$, then the maximum local value of the wave amplitude η_{max} is given by

$$\frac{\eta_{max}}{H} = \left(\frac{X_0}{x}\right)^{1/4} = \frac{1}{h^{1/4}}, \tag{19}$$

where $x = h \cot\beta = hX_0$; this is the familiar evolution referred to as Green's law. This appears as the first time that Green's law had been shown to be valid on the combined topography and including reflection.

2.6. *Validity of the Solution*

The linear theory and nonlinear theory solutions described are valid for functions $\Phi(k)$ such that the Jacobian of the transformation (5) is never equal to zero. The Jacobian becomes zero when the surface slope $\partial\eta/\partial x$ becomes infinite. In the physical plane, this point is usually interpreted as the point of wave breaking.

Extending Synolakis's (1987) analysis, and taking the limit Jacobian of the Carrier and Greenspan transformation as $\sigma \to 0$, then it can be shown that the Jacobian is always regular when

$$u_\lambda - \tfrac{1}{2} = -\tfrac{2}{3}X_0^2 \int_{-\infty}^{\infty} \frac{\kappa^2 \Phi(k)\, e^{i\kappa(X_1 - X_0 + (\lambda/2)X_0)}}{J_0(2\kappa X_0) - iJ_1(2\kappa X_0)}\, d\kappa - \tfrac{1}{2} \neq 0. \tag{20}$$

The calculation of the runup integral should always be accompanied by the evaluation of the Jacobian to ensure that there is no wave breaking; again once $\Phi(k)$ is known, this is a straightforward numerical evaluation. It should also be noted (Tuck and Hwang, 1972; Synolakis, 1987) that this formalism is so robust that frequently realistic results can be obtained with the rump integral for waves for which the Jacobian does go through zero. The physical explanation is that the shallow-water wave equations may predict breaking for waves that do not actually break in nature. However, extreme caution should be exercised when interpreting results obtained when (20) is violated.

3. Comparisons with Laboratory Data and Higher-Order Theories

The real test of the linear theory is in the laboratory. One could argue that the runup invariance is an artifact of the particular solution method used and is not therefore generally applicable. The data suggest that for linear theory models, the laboratory results are within the same small variance as higher-order theories.

Consider the runup of solitary waves. Although it has long been known that breaking and nonbreaking periodic waves follow different runup variations, this behavior has never been recognized in single-wave runup. In Synolakis (1987), data was presented that conclusively demonstrated that two different runup regimes exist, one for breaking and one for nonbreaking solitary waves. As a result, comparisons of analytical results with laboratory data such as those of Hall and Watts (1953) can no longer be considered as straightforward as often assumed. That study includes both breaking and nonbreaking wave data without identifying them as such, and therefore the empirical relationships derived in that study may not be directly applicable when determining the runup of nonbreaking waves. To perform a posteriori identification of the Hall and Watts data, the breaking criterion

$$H > 0.818(\cot\beta)^{-10/9} \tag{21}$$

derived from Equation (20) for solitary waves was used, and the identified non-breaking wave data are presented in Figure 2. The abscissa is the runup law (Equation (14)), and the ordinate is the maximum runup. The asymptotic result derived from the linear theory solution does appear to model the laboratory data quite well.

All the data in Figure 2 are from the Hall and Watts study, with the exception of the 1 : 2.75 data (Pedersen and Gjevik, 1983) and the 1 : 20 data (Synolakis, 1986).

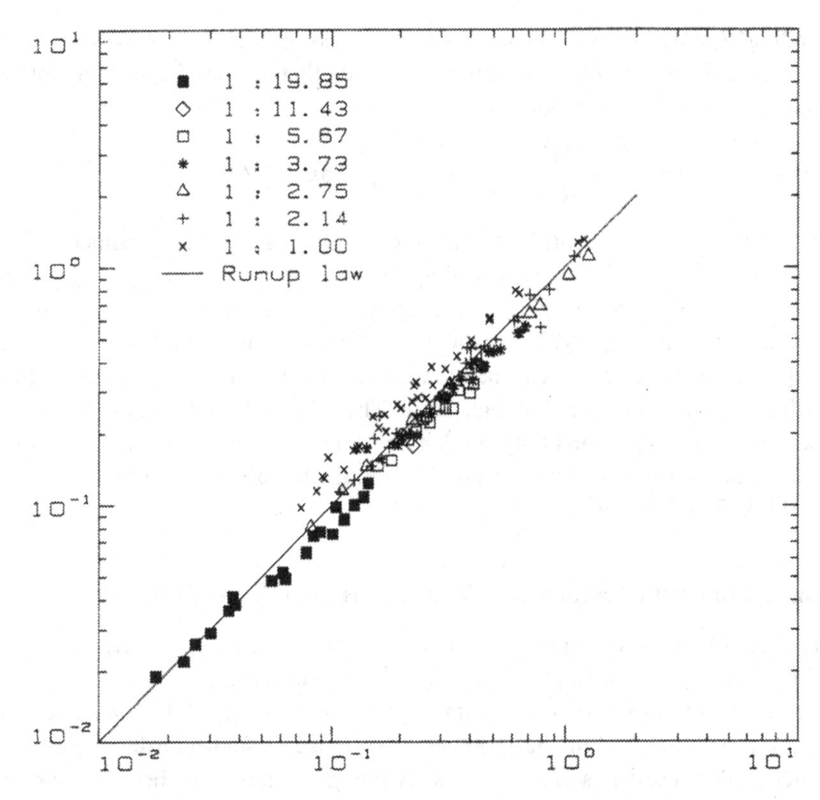

Fig. 2. The maximum runup **R** of solitary waves of height H up a beach of slope β. Comparison of the runup law (Equation (14)) with laboratory data from Hall and Watts (1955), and Pedersen and Gjevik (1983) (slope 1 : 2.75) and Synolakis (1986) (slope 1 : 19.85).

The latter were generated in a laboratory wave channel equipped with a programmable hydraulic-wave generator with the procedure described in Synolakis (1990). This procedure has repeatedly been shown to accurately reproduce Boussinesq-type solitary waves. The fact that maximum runup data generated with relatively primitive methods such as those of Hall and Watts agree so well with data generated using very elaborate methods, suggests that the maximum runup – to first order – is relatively insensitive to small details in the wave profile and to the roughness of the laboratory channel.

To evaluate the differences in the predictions of the linear theory and of the nonlinear-dispersive theory consider table I, which compares the predictions of the runup law with the predictions of Boussinesq-type numerical solutions. Table II is a comparison of the predictions of the runup integral with the state-of-the-arts boundary element solution of Grilli and Svendsen (1989) who, without depth-averaging solve, the horizontal velocity field. It is again seen that linear theory models the laboratory data equally well and sometimes even better than higher-order theories.

Table I. Solitary wave runup results from numerical calculations using one-dimensional Boussinesq type models. H&H refers to Heitner and Housner (1970), KLL to Kim, Liu and Ligett (1983) and P&G to Pedersen and Gjevik (1983). The results are compared to the predictions of the runup law (Equation (13)) and with the laboratory data of Hall and Watts (1955), when possible.

Source	cot β	H/d	$\mathbf{R}/d$ Numerical calculations	$\mathbf{R}/d$ Runup law $2.8X_0^{1/2}(H/d)^{5/4}$	$\mathbf{R}/d$ Laboratory experiments
H&H	10.000	0.030	0.100	0.112	na
H&H	10.000	0.050	0.180	0.212	na
KLL	3.732	0.050	0.135	0.129	0.173
KLL	3.732	0.100	0.308	0.308	0.281
KLL	3.732	0.200	0.766	0.732	0.599
H&H	3.333	0.050	0.150	0.122	na
H&H	3.333	0.100	0.310	0.291	na
P&G	2.747	0.050	0.127	0.111	0.115
P&G	2.747	0.098	0.275	0.257	0.252
P&G	2.747	0.193	0.599	0.600	0.552
P&G	2.747	0.294	0.958	1.016	0.898
KLL	1.000	0.060	0.129	0.084	0.115
KLL	1.000	0.100	0.159	0.159	0.212
KLL	1.000	0.200	0.504	0.379	0.454
KLL	1.000	0.480	1.610	1.131	1.270

Table II. Solitary wave runup results from Grilli and Svendsen (1989) using a two-dimensional potential flow model. The results are compared to the predictions of the runup integral (Equations (8)).

Comparison of BEM results with Synolakis (1987)

Slope angle	H	$\mathbf{R}$ (BEM)	$\mathbf{R}$ (Synolakis)
45°	0.269	0.67	0.644
45°	0.457	1.28	1.194
70°	0.269	0.58	0.544
70°	0.457	1.07	0.951

Figure 3 shows the maximum normalized runup of cnoidal waves $\mathbf{R}$ with the same period $T = 16$ as a function of the dimensionless wave height H up three different sloping beaches. The numerical calculations of Ohayama (1986) are compared with the linear theory (Equations (15)), with data for the runup of the equivalent sinusoidal wave and with laboratory data. ('Equivalent' sinusoidal and cnoidal waves have the same period and wave height). The figure suggests that the runup of a cnoidal wave is substantially higher than the runup of the monochromatic wave with the same height and wavelength. Also, it is seen that linear theory

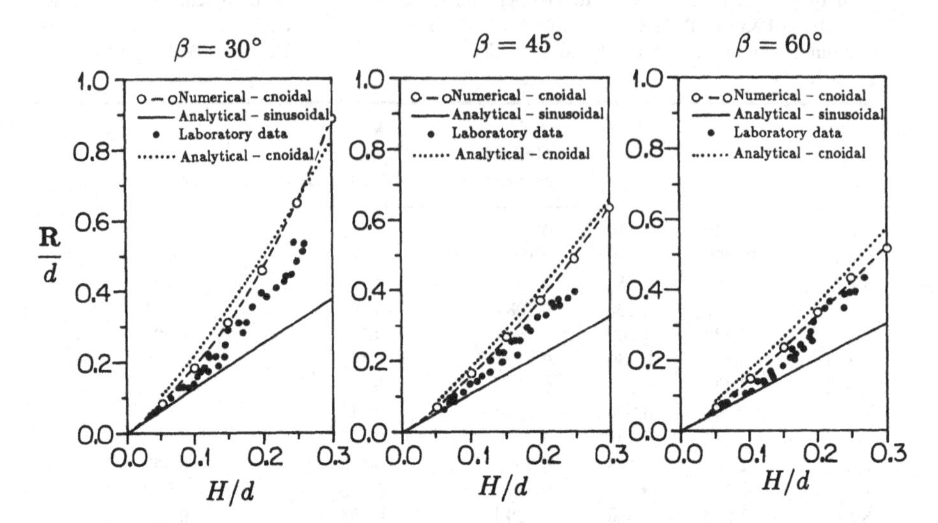

Fig. 3. The maximum runup **R** of cnoidal waves of height H and period 16 up three different sloping beaches. Laboratory data are compared with Ohayama's (1986) solution – referred to as numerical cnoidal – with the predictions of Equation (15) – referred to as analytical cnoidal – and with the Keller and Keller (1964) sinusoidal wave solution – referred to as analytical sinusoidal. Notice that the maximum runup of any cnoidal wave in the figure is significantly higher than the maximum runup of the sinusoidal wave with the same height and period.

predictions agree well with Ohayama's numerical results which are derived with boundary integral methods.

These results are valid for waves climbing up steep beaches or for very long waves. Most natural beaches are gently sloping, and it is not obvious if similar differences exist. Interestingly, Figure 3 does suggest that a common practice of choosing the dominant frequency of an incoming wave-spectrum and using the runup of the monochromatic wave with that dominant frequency for calculating wave runup, may not be entirely appropriate when the incoming wave energy is dominated by low frequency swell (Synolakis, 1988a).

Figure 4 shows results from the evolution of the maximum amplitude of three different solitary waves up a 1 : 20 sloping beach; the $H = 0.3$ is a breaking wave. The solid lines through the three sets of data represent the relationship $\eta_{\max} \sim h^{-1/4}$, i.e. Green's Law. It is clear that linear theory predicts the evolution of the maximum amplitude satisfactorily, even for breaking waves.

4. Conclusions

In this paper, a method has been presented for solving both the linear and nonlinear forms of the shallow-water wave equations. It has been shown that linear

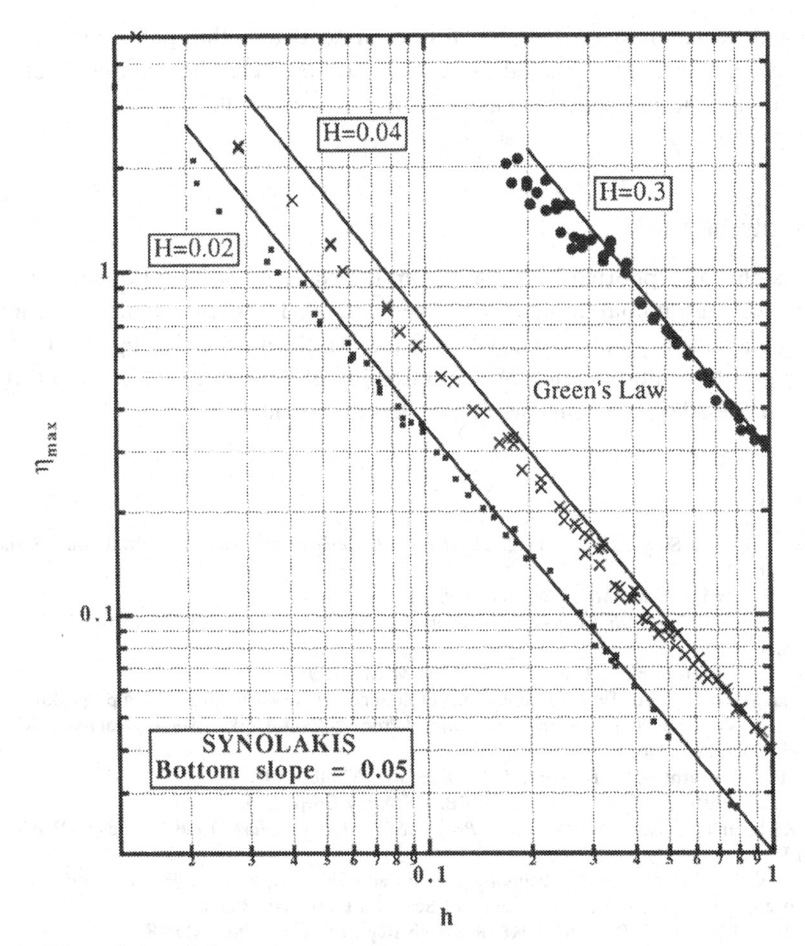

Fig. 4. The evolution of the maximum amplitude of three solitary waves on a 1:20 beach. The solid line is Green's Law (Equation (19)). Data from Synolakis (1986).

and nonlinear theory predict mathematically identical maximum runup heights. Linear theory has been shown to predict Green's law for the evolution of the maximum amplitude of solitary waves over the combined topography and including reflection. A comparison with laboratory data has been presented which demonstrates that linear theory can be used effectively to calculate the maximum runup of tsunamis and the maximum height evolution. The calculation involves the specification of the transform of the incoming tsunami and then the direct evaluation of the maximum value of the runup integral (Equation (8)). It is much simpler than numerically solving higher-order models; however, runup calculation should always be accompanied by calculation of the Jacobian of the Carrier and Greenspan transformation (Equation (20)) to ensure that there is no wave breaking.

It should be emphasized that the results presented in this paper are only of engineering interest; even though linear theory predicts the evolution of the maximum wave height and the maximum runup quite well, it does not correctly model the details of the profile evolution.

Acknowledgements

This research was supported by a grant from the National Science Foundation No. 53-4507-5909. Additional funding has been provided through another National Science Foundation grant (managed by Fred Raichlen of CIT), by the Faculty Research and Innovations Fund of the University of Southern California and by a grant from the Alexander Onassis Public Benefit Foundation.

References

Abramowitz, M. and Stegun, I. A.: 1972, *Handbook of Mathematical Functions*, Natl. Bur. Stands., Washington, D.C.

Carrier, G. F.: 1966, *J. Fluid Mech.* **24**, 641–659.

Carrier, G. F.: 1971, *Proc. 6th Summer Seminar on Applied Mathematics*, RPI, Troy, NY, 1970, Amer Math. Soc.

Carrier, G. F. and Greenspan, H. P.: 1958, *J. Fluid Mech.* **17**, 97–110.

Grilli, S. and Svendsen, I. A.: 1989, Computations of nonlinear wave kinematics during propagation and runup on a slope, in *Water Wave Kinematics* (Proc. NATO-ARW, Molde, Norway), Kluwer Academic Publishers, Dordrecht.

Guza, R. T. and Thornton, E. B.: 1982, *J. Geophys. Res.* **87**, 483–491.

Hall, J. V. and Watts, J. W.: 1953, TM 33, BEB, US Army Corps Eng.

Heitner, K. L. and Housner, G. W.: 1970, *Proc. ASCE, J. Wat. Harb. Coastal Engng.*, **WW3**, pp. 701–719.

Kaistrenko, V. M., Mazova, R. K., Pelinofsky, E. N. and Simonov, K. V.: 1985, *Tsunami Runup on Shore*, Inst. Appl. Phys., U.S.S.R. Academy of Sciences, Gorky (in Russian).

Keller, J. B. and Keller, H. B.: 1964, ONR Research Report Contract No. NONR-3828(00), Dept. of the Navy, Washington, D.C.

Kim, S. K., Liu, P. L-F. and Ligett, J. A.: 1983. *Coastal Engng.* **7**, 299–317.

Lamb, H.: 1932, *Hydrodynamics*, 6th edn., Cambridge University Press.

Pedersen, G. and Gjevik, B.: 1983, *J. Fluid Mech.* **135**, 283–299.

Pelinofsky, E. N.: 1989, *Sci. Tsun. Hazards* **7**, 117–126.

Pelinofsky, E. N., Golinko, V. I. and Mazova, R. K.: 1989, Tsunami wave runup on a beach; Exact analytical results, preprint No. 232, Inst. Appl. Phys., U.S.S.R. Academy of Sciences, Gorky (in English).

Ohyama, T.: 1987, *Proc. JSCE* **381**, II-7, 189–198. (in Japanese.)

Shuto, N.: 1973, *Coastal Engineering in Japan* **16**, 25–42.

Synolakis, C. E.: 1986, The runup of long waves, PhD Thesis, California Institute of Technology, Pasadena.

Synolakis, C. E.: 1987, *J. Fluid Mech.* **185**, 523–545.

Synolakis, C. E., Deb M. K. and Skjelbreia, E. J.: 1988a, *Phys. Fluids* **31**, 1–4.

Synolakis, C. E.: 1988b, *Quart. Appl. Math.* **46**, 105–107.

Synolakis, C. E.: 1990, *J. Water. Harb. Coast. Eng.* **116**, 252–266.

Svendsen, I. A.: 1974, Cnoidal waves over gently sloping bottom. Inst. Hydr. Eng., Techn. Univ. Denmark, Ser. Paper 6, Lyngby, Denmark.

Tuck, E. O. and Hwang, L.: 1972, *J. Fluid Mech.* **51**, 449–461.

Yeh, H.: 1991, Tsunami bore run-up, *Natural Hazards* **4**, 209–220 (this issue).

Natural Hazards 4: 235–255, 1991.

A Numerical Model for Far-Field Tsunamis and Its Application to Predict Damages Done to Aquaculture

OSAMI NAGANO,
Electric Power Development Co. Ltd., Chigasaki, Kanagawa 253, Japan

FUMIHIKO IMAMURA and NOBUO SHUTO
Department of Civil Engineering, Tohoku University, Aoba, Densai 980, Japan

(Received: 15 March 1990; revised: 1 June 1990)

Abstract. The 1960 Chilean tsunami which traveled the Pacific Ocean and caused much damages to Japan is simulated from its generation to the terminal effects on coastal areas. In the computation of ocean propagation by the linear longwave theory, a new technique is introduced to keep the same accuracy as the linear Boussinesq equation and reduce the CPU time as well as the computer memory. In the coastal transformation computation, the energy dissipation due to sea-bottom scouring is suggested to be included, particularly in the case of long bays. To obtain accurate results, the current velocity requires finer spatial grids than the water surface elevation. Damage done to pearl culture rafts are explained in terms of the computed current velocity.

Key words. The 1960 Chilean tsunami, numerical simulation, Imamura number, difference between far- and near-field tsunamis, sea-bottom scouring, damage to aquaculture.

1. Introduction

In the early morning on 24 May 1960, the whole Japanese coastal area on the Pacific Ocean was surprised by an unanticipated attack of a tsunami. The tsunami was 5–6 m high along the Sanriku coast and 3–4 m at other coastal areas. Major damage was caused to the south coast of Hokkaido, the Sanriku coast, and the Kii Peninsula. The tsunami claimed 119 lives with 20 people missing. More than 2800 houses were destroyed and washed away and 2183 houses were partially destroyed. Aquacultures (pearl and oyster cultures) were heavily damaged.

The tsunami was generated off the Chilean coast on the day before, traveled over the Pacific Ocean, and converged on the Japanese Archipelago. This experience opened the way for modern tsunami prevention work in Japan, the main method of which was to heighten the existing sea walls and to build new sea walls and tsunami breakwaters. The trace height of this tsunami was taken as the design height of the sea walls. For the design of the tsunami breakwaters at Ofunato Bay, a tsunami wave profile was needed, because the breakwaters reduced the tsunami height by changing the oscillation characteristics of the bay. Due to a lack of detailed tsunami profile data, the breakwaters were designed by assuming that the tsunami consisted of sinusoidal waves of constant amplitude and period.

Just after the Chilean tsunami, several attempts were made to simulate it numerically. The wave-ray method was only successful in simulating the refraction and travel time of the tsunami. Other attempts ended in failure. The most vital problem was the lack of a method of estimating the initial profile. This was solved when Mansinha and Smylie introduced a method in 1971. Another difficulty was found in numerical simulations of a transoceanic propagation of tsunamis. If one wished a high accuracy, a huge number of spatial grids was required. This was not handled well until supercomputers appeared.

The present paper aims to simulate the Chilean tsunami by using the recent numerical techniques mostly developed for nearfield tsunamis.

In the case of a near-field tsunami, a tsunami in the past can be simulated within a 15% error as far as the maximum runup height and the area flooded are concerned. In the case of a far-field tsunami, the number of grid points becomes huge, thereby providing an opportunity for the introduction and accumulation of numerical errors. No example has ever been proposed to discuss the accuracy of computed results.

In Section 2, the equations and conditions are discussed. The computation is divided into two: ocean propagation and coastal transformation computations. In the former, the linear longwave equations with Coriolis force in the longitude-latitude coordinates are used. When discretized, the Imamura number is set nearly equal to unity, thereby introducing the numerical dispersion in place of the physical dispersion which must be included in the propagation problem over the ocean. Outputs of the ocean propagation computation is inputted at the sea boundary of the coast transformation computation. The coast region is again divided into two: offshore and near-shore regions. In the offshore region, the linear longwave theory without Coriolis force in the Cartesian coordinates is used. In the near-shore region, the shallow-water theory with bottom friction is used.

In Section 3, the ocean propagation problem is solved and the overall feature of the tsunami in the ocean and along the Japanese Archipelago is compared with tidal records.

In Section 4, the coast transformation problem is solved. The difference between near- and far-field tsunami is shown for the spatial distribution of the maximum elevation of the water surface. Detailed examinations reveal that the computation cannot simulate the tsunami in several long bays where severe bottom scouring occurred.

In Section 5, the accuracy of the current velocity computation is discussed. Damage done to pearl culture rafts are explained as a function of the computed current velocity and water level.

2. A Numerical Model for Far-Field Tsunamis

2.1. *The Ocean Propagation Model*

According to Kajiura's criterion (1963), the propagation of a tsunami over the Pacific Ocean should be solved using the linear Boussinesq equation which includes dispersion term. Imamura *et al.* (1988) showed that the 'staggered leap-frog' scheme inevitably introduced the numerical dispersion term into the difference equations for linear long waves, and that this numerical dispersion term could replace the physical dispersion term in the linear Boussinesq equation, if the grid length was appropriately selected; that is, the following Imamura number (see Appendix),

$$\mathrm{Im} = \Delta x[1 - (C_0 \Delta t/\Delta x)^2]^{1/2}/2h \tag{1}$$

should be taken to be nearly equal to unity. Here, Δx and Δt are the spatial and temporal grid lengths, C_0 the celerity of linear long waves, and h the water depth. It was 1.56 in the present simulation, the spatial grid length being $10'$ ($=20$ km, on average). The time increment was determined to satisfy the CFL condition. This consideration saved the CPU time and computer memory, still keeping the same accuracy as the linear Boussinesq equation.

The following equations were, therefore, used.

$$\begin{aligned} &\frac{\partial \eta}{\partial t} + \frac{1}{R \cos \theta}\left\{\frac{\partial M}{\partial \lambda} + \frac{\partial}{\partial \theta}(N \cos \theta)\right\} = 0, \\ &\frac{\partial M}{\partial t} + \frac{gh}{R \cos \theta}\frac{\partial \eta}{\partial \lambda} - fN = 0, \\ &\frac{\partial N}{\partial t} + \frac{gh}{R}\frac{\partial \eta}{\partial \theta} + fM = 0 \end{aligned} \tag{2}$$

in which η is the water surface elevation, M and N discharge fluxes along the latitude θ and longitude λ, h the water depth, f the Coriolis parameter, g the gravitational acceleration, and R the radius of the Earth.

The smallest water depth was taken to be 200 m. Perfect reflection was assumed at the land boundary. The initial profile of the 1960 Chilean tsunami is shown in Figure 1, computed from the fault parameters obtained by Kanamori and Ciper (1974).

2.2. *The Coast Transformation Model*

The coast transformation model consisted of two computations: offshore and near-shore computations. The linear longwave theory (Equation (3)) was used in the offshore region deeper than 50 m, and the shallow-water theory with bottom friction (Equation (4)) in the near-shore region shallower than 50 m.

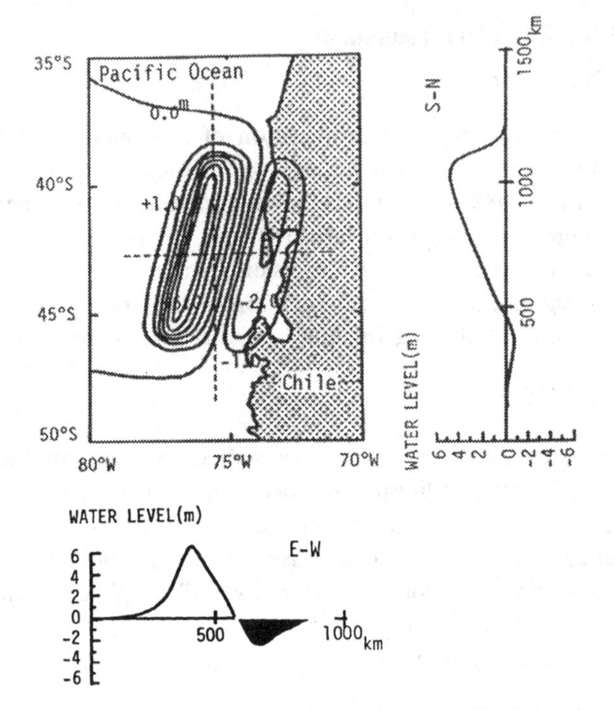

Fig. 1. Initial profile of the 1960 Chilean tsunami.

$$\frac{\partial \eta}{\partial t} + \frac{\partial M}{\partial x} + \frac{\partial N}{\partial y} = 0,$$

$$\frac{\partial M}{\partial t} + gh\frac{\partial \eta}{\partial x} = 0, \tag{3}$$

$$\frac{\partial N}{\partial t} + gh\frac{\partial \eta}{\partial y} = 0.$$

in which M and N are discharge fluxes in the x and y directions.

$$\frac{\partial \eta}{\partial t} + \frac{\partial M}{\partial x} + \frac{\partial N}{\partial y} = 0$$

$$\frac{\partial M}{\partial t} + \frac{\partial}{\partial x}\left(\frac{M^2}{D}\right) + \frac{\partial}{\partial y}\left(\frac{MN}{D}\right) + gD\frac{\partial \eta}{\partial x} + gn^2\frac{MQ}{D^{7/3}} = 0, \tag{4}$$

$$\frac{\partial N}{\partial t} + \frac{\partial}{\partial x}\left(\frac{MN}{D}\right) + \frac{\partial}{\partial y}\left(\frac{N^2}{D}\right) + gD\frac{\partial \eta}{\partial y} + gn^2\frac{NQ}{D^{7/3}} = 0,$$

in which Q is the resultant discharge flux, $D = h + \eta$ the total water depth, and n Manning's roughness.

Outputs of the ocean propagation computation at the water depth of about 1000 m along the Japanese Archipelago were used as the sea-boundary conditions of the coastal transformation computation. At the land boundary, perfect reflection was assumed, because our major concern was not tsunami runups. It means that no moving boundary was taken into account. The CPU time was thus very much reduced.

The grid length was carefully determined. To eliminate the decay of wave height due to discretization, a local wavelength must be covered by at least 20 grid points (Shuto *et al.*, 1986). The time increment was determined to satisfy the CFL condition.

3. Results of the Ocean Propagation Computation

3.1. *Distribution in the Pacific Ocean*

The propagation diagram of the tsunami in the ocean is shown in Figure 2. Contours are the position of the tsunami front at 1 hr. time intervals and the attached numerals are the time after the earthquake. Solid circles in the figure are the tide gauge stations, and the attached numerals are the maximum recorded tsunami wave height in meters. During the first four hours, the tsunami front propagates as concentric circles. Then it becomes nearly straight parallels whose

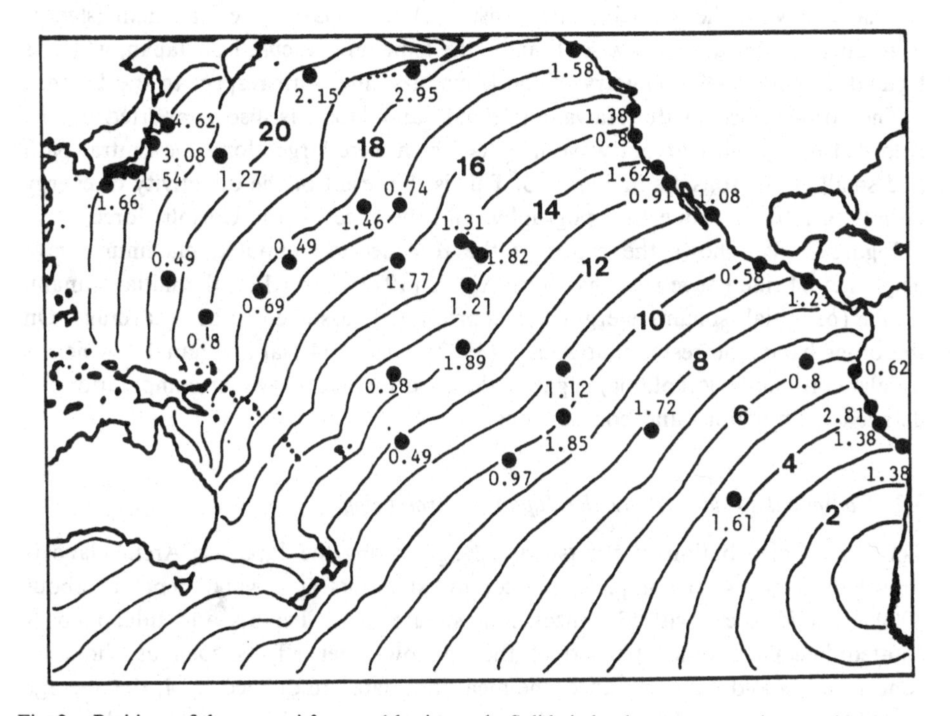

Fig. 2. Positions of the tsunami front at 1 hr. intervals. Solid circles denote measured tsunami in meter.

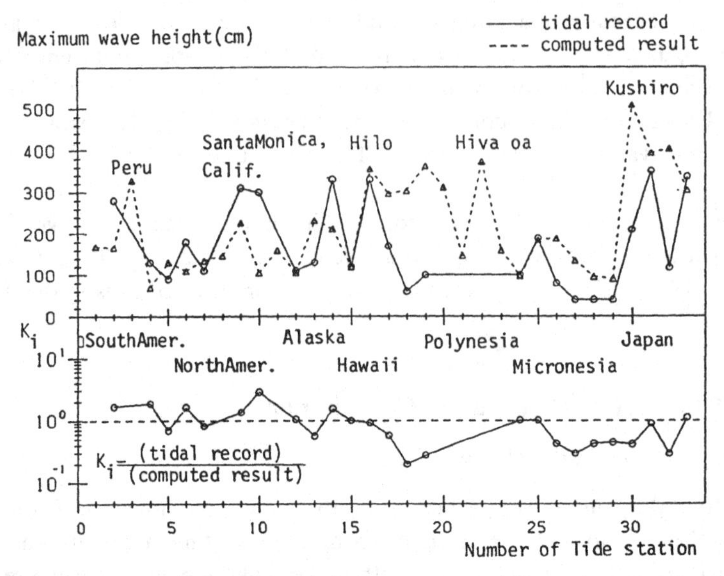

Fig. 3. (a) Comparison of the measured data (solid lines) with outputs (dotted lines) of the ocean propagation computation in the Pacific Ocean. (b) Ratios of the recorded-to-computed maximum wave heights.

normal is toward the Kamchatka Peninsula. After it passes the Hawaiian Islands, it gradulally turns anticlockwise. Finally, the tsunami is focused on Japan, which is located at a point of symmetry of the tsunami source with respect to the Earth.

The convergence to the Hawaiian Islands and Japan is also supported by the recorded maximum tsunami wave heights which were large along the central path and small at environs, as in Figure 3. This is the result of the directivity of energy radiation, refraction due to topography, and the effect of the Coriolis force.

Figure 3 also shows the ratios of the recorded-to-computed maximum wave heights at 34 tide gauge stations. Since Aida's (1977) K is 1.121, the initial tsunami profile (or initial tsunami energy) was appropriately assumed to be an average. On the other hand, the result that Aida's (1977) κ is 1.934 (larger than 1.4 which is usually considered acceptable) suggests the need for shallow-water computation for detailed examinations and comparisons.

3.2. *Computed Results Along the Japanese Archipelago*

The overall distribution of the tsunami height along the Japanese Archipelago is shown in Figure 4. It compares the computed results at a water depth of about 3000 m (solid lines) with the measured tsunami trace heights and tidal records (dotted lines). In order to correct the shallow-water effects such as shoaling, concentration and bay-resonance, the measured data are divided by 4, the average amplification factor. The computed and measured tsunami heights show the same

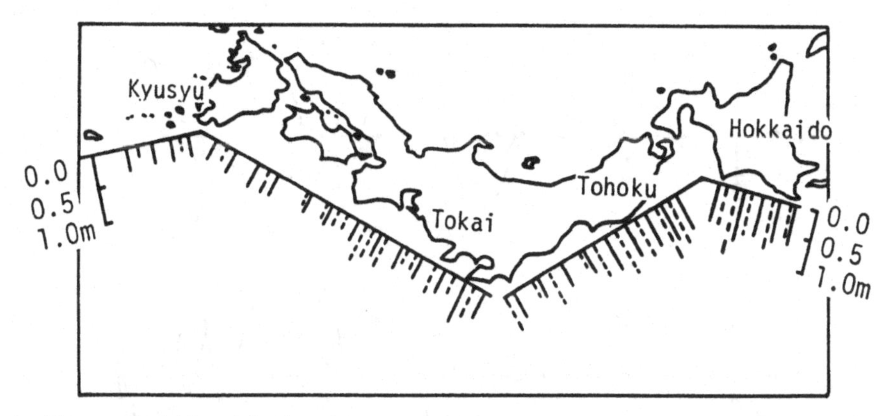

Fig. 4. The measured (dotted lines) and computed (solid lines) tsunami heights along Japan (ocean propagation computation).

tendency that the tsunami is higher at the Hokkaido and Tohoku districts, and smaller to the west of the Tokai district.

Comparisons in detail are given in Figure 5, which shows the time histories of the tsunami obtained in the ocean propagation computation (thin solid lines) and the tsunami recorded on tide gauges or by eye measurements (dotted lines). In the ocean propagation problem, the shallowest water depth near the shoreline was assumed to be 200 m, and the actual topography was much simplified because of the wide spatial grid of 10′. The local topographical effects (shoaling, concentration and resonance) were, therefore, not fully taken into account. In spite of these limitations, agreements between the computed and measured are reasonably good, except for at Owase which is located in the middle of the Kii Peninsula, a typical ria coast. The general features of the Chilean tsunami as well as its major wave period are well simulated: the small first rise of the water level followed by a large fall. Further examination and improvement are necessary for the magnitude of amplitude and the profile in the later stage, by using the shallow-water theory and finer spatial grids. This will be discussed again in Section 4.1.

3.3. *The CPU Time*

Table I summarizes the computation condition and the CPU time. The supercomputer used was an SX-1 of the NEC corporation. The 42 hr. tsunami over the Pacific Ocean was reproduced within 12 min 17 sec with the linear longwave equation and within 45 min 36 sec by the linear Boussinesq equation. The former was solved with an explicit scheme and the latter with an implicit scheme. This shows the advantage of the use of the linear longwave equation under the condition that the Imamura number is set nearly equal to unity.

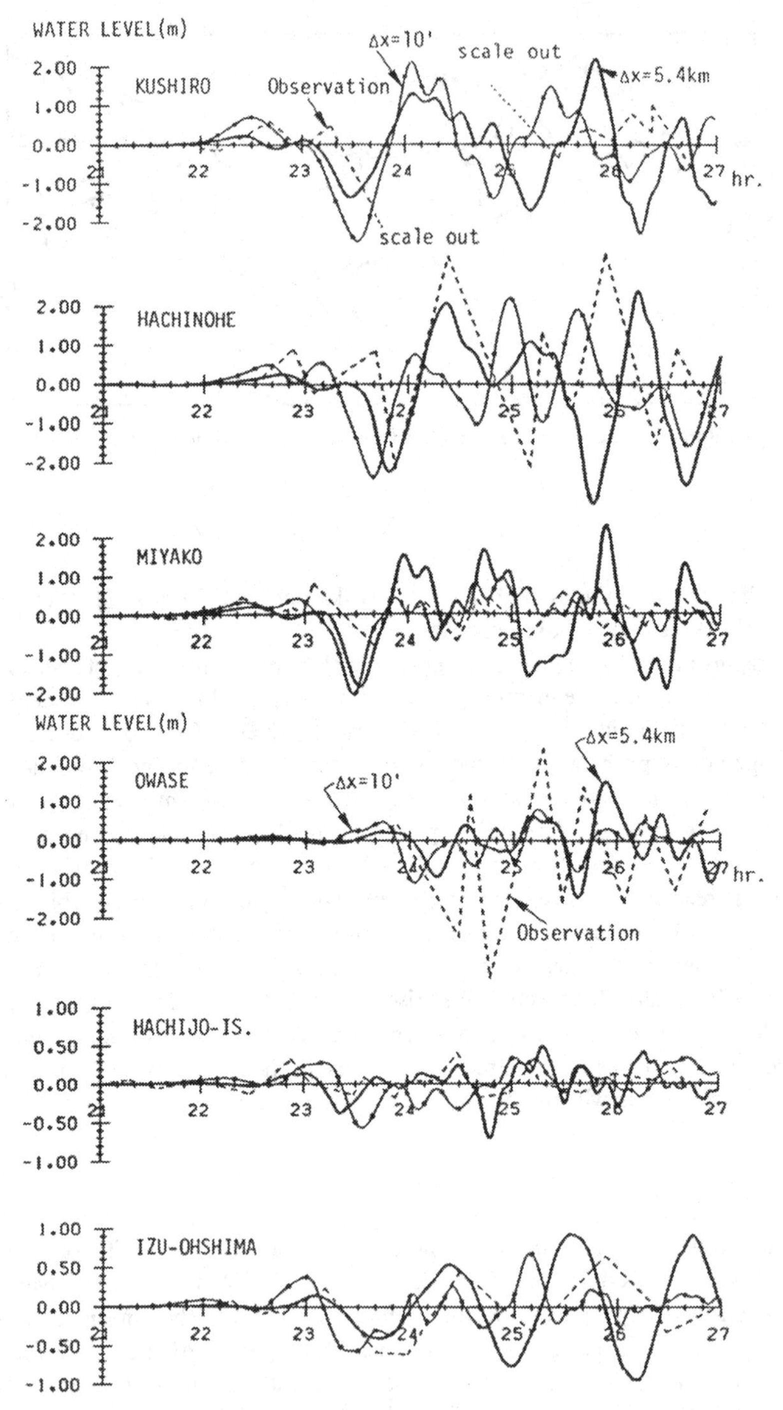

Fig. 5. Time histories of the water level (thin lines: ocean propagation computation, solid lines: offshore computation, dotted lines: tide records).

Table I. Equations, computation conditions, and CPU times

Equation	Linear Boussinesq Equation	Linear longwave theory
Grid length	10′	
Number of Grids	734 400	
Reproduction time	42 hr	
CPU time	45 min 36 sec	12 min 17 sec
(ratio)	(3.71)	(1.00)
Computation size	45 MByte	28 MByte

4. Results of the Coastal Transformation Computation

4.1. *Tsunami Waves Along Japan*

The first stage of the coastal transformation problem was the offshore computation in which the grid length was 5.4 km and the shallowest water depth was 200 m.

The time histories of the water-surface elevation obtained in the offshore computation are shown by thick solid lines in Figure 5. Compared to the ocean propagation computation (thin lines), the results are improved except at Owase, the complicated bay shape of which had not yet been fully taken into account even in the offshore computation. For better agreements, the spatial grid length should be made finer as in the following near-shore computation.

4.2. *Results fo the Near-Shore Computation*

The Sanriku coast and the Kii Peninsula were selected for the near-shore computation. The linear longwave theory was used with coarse grids ($\Delta x = 5.4$ km) in deep sea. In shallow sea, the shallow-water theory with bottom friction was used with grids which were made finer towards land. The finest grid length was 200 m. No runup was taken into account at the shore. The outputs of the offshore computation were used as the sea-boundary conditions.

Figures 6(a) and (b) are examples of the computed results. Contours of the computed maximum water level are shown in Figure 6(a). The computed tsunami heights (solid lines) are compared with the recorded data (open circles and/or bars which give the range of the measured data) in Figure 6(b). Lines with dots are the results of the linear longwave theory obtained with fine grids. Although the nonlinear and bottom friction terms improve the results, agreement is not good at the bottom of long bays. Outside bays, agreement is reasonably good.

4.3. *Difference Between Far- and Near-Field Tsunamis*

The 1896 Meiji Great Sanriku tsunami which was the most disastrous tsunamis in Japan during the past two centuries, was selected as the near-field tsunami for

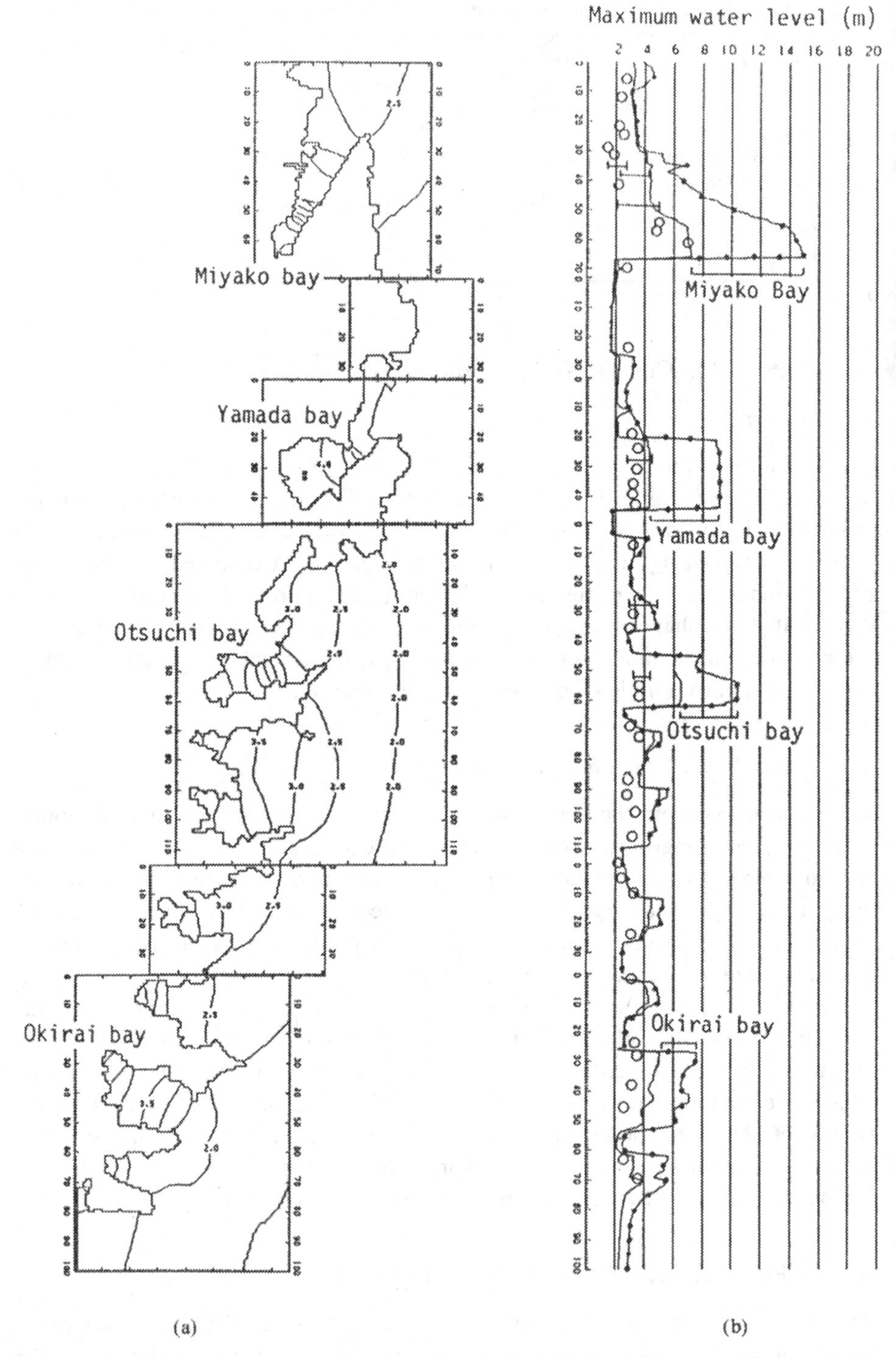

Fig. 6. (a) Contours of the computed maximum water level by the shallow-water theory, and (b) Comparison of the measured tsunami heights (open circles and bars) and the computed in the near-shore computation with the linear (lines with dots) and shallow-water (lines) theories.

comparison. In the case of a near-field tsunami, contours of the computed maximum water level are parallel to the coastline as shown in Figure 7(a). This suggests that shoaling and converging are the major amplification mechanisms. On the other hand, in the case of a far-field tsunami, contours are normal to the bay axis, as shown in Figure 7(b), suggesting that resonance in bays is the dominant amplification mechanism.

4.4. *The Chilean Tsunami in Bays*

When a tsunami enters a bay, three waves in sequence are necessary and sufficient for full resonance of the bay water (Kajiura, 1963). Figure 8(a) shows the time histories of the water-surface elevation at three points in Kesennuma Bay given by solid circles in Figure 8(b), computed with the linear (lines with dots) and shallow water (solid lines) theories. Since more than three waves enter the bay, the resonance is complete. Figure 8(c) shows comparisons of the computed and measured tsunami heights (open circles and bars). The linear theory (lines with

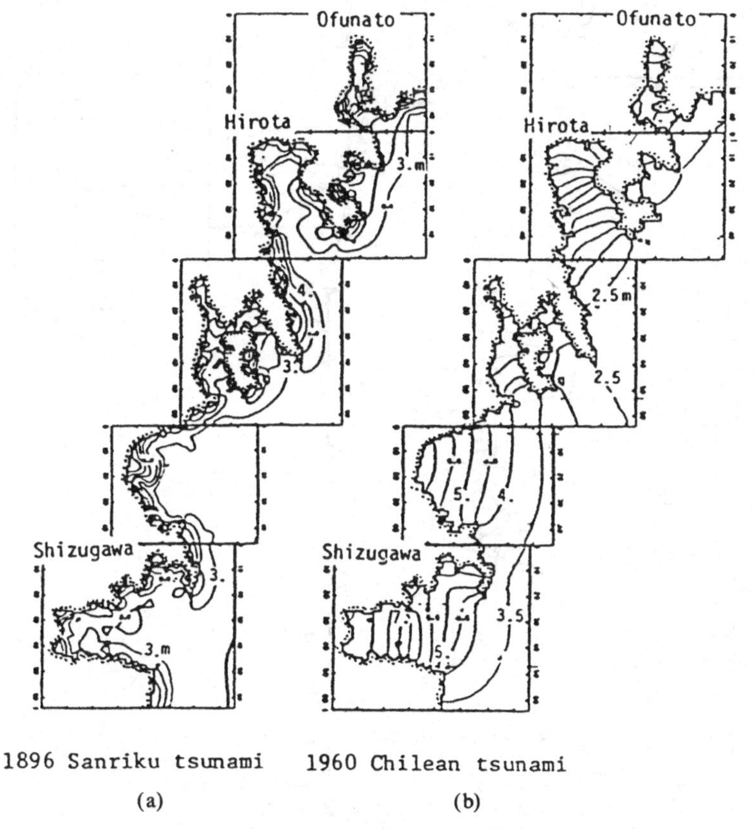

Fig. 7. Computed maximum water levels of the 1896 Meiji Great Sanriku tsunami and the 1960 Chilean tsunami.

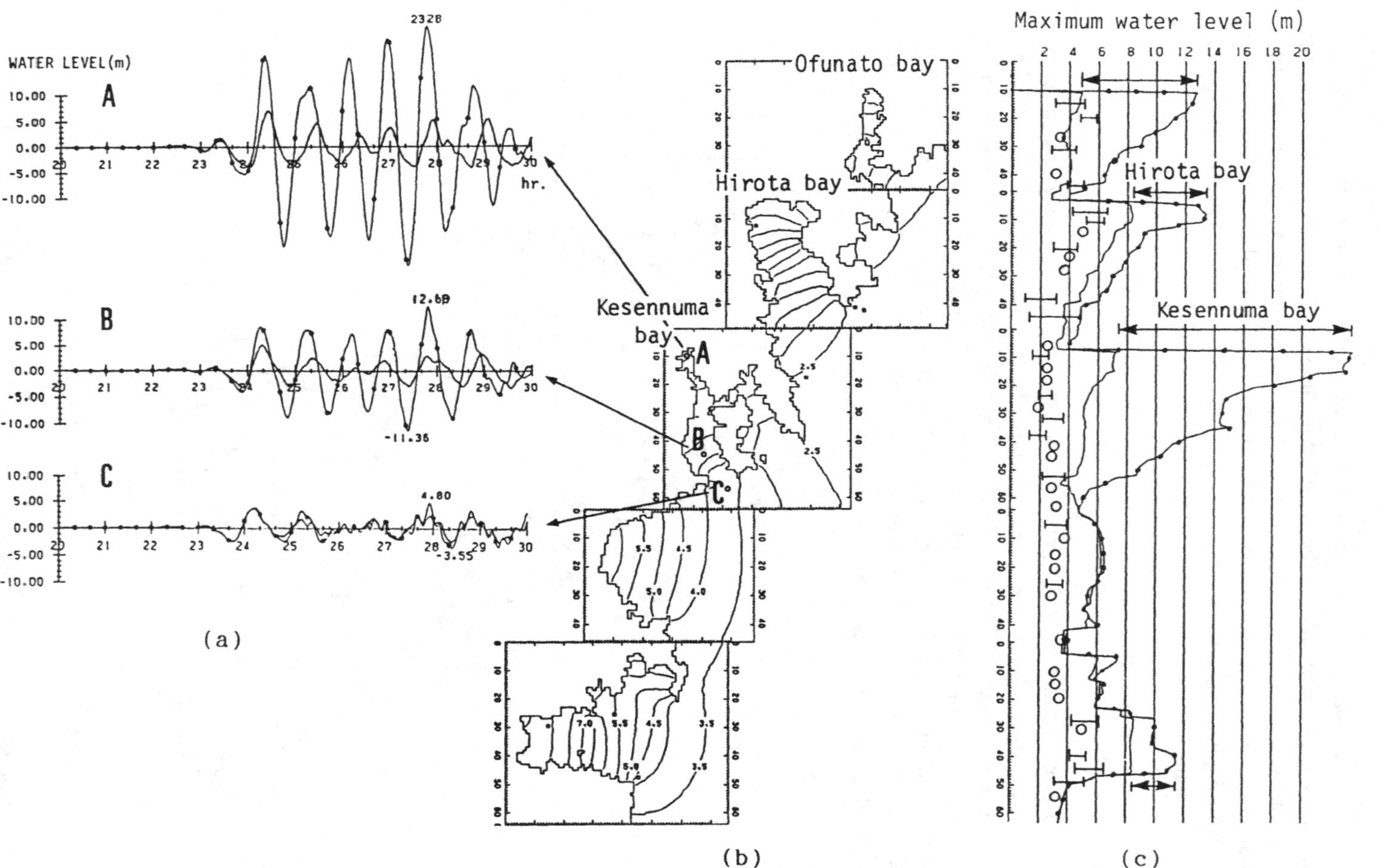

Fig. 8. Comparisons in detail. (a) Time histories in Kesennuma Bay with the linear (lines with dots) and shallow-water (solid lines) theories, (b) Contours of the maximum water level obtained by the shallow–water theory, and (c) Comparison of the computed tsunami heights with the measured (open circles and bars).

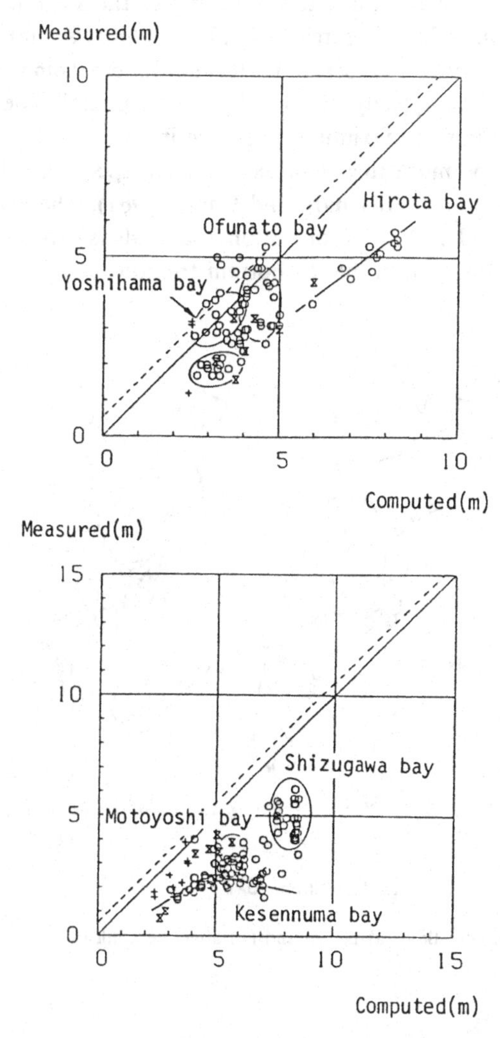

Fig. 9. Comparisons of the computed and measured tsunami heights in bays. Open circles and crosses are for the inside and outside of bays. The underlined are long bays.

dots) gives very large values in the bay, because it does not include the bottom friction term. The shallow-water theory (solid lines) gives better results because of the bottom friction and nonlinear terms. The computed tsunami height with the shallow-water theory is, however, still much greater than the measured data at the bottom of the bay.

Another comparison between the measured and computed tsunami heights in bays in the Sanriku district is shown in Figure 9. The worst agreement is obtained in longer bays. This suggests the need for other energy dissipation mechanisms than bottom friction.

The answer may be the energy dissipation due to a scouring of the sea bottom, as surveyed in the case of Kesennuma Bay. Figures 10(a) and (b) (Kawamura and Mogi, 1961) show the sea bottom contours before and after the Chilean tsunami. A jetty in the middle of the bay was completely destroyed by the tsunami. The sea bottom was severely deepened by 10 m at maximum, as shown in Figure 10(c). The current velocity was estimated to be more than 6 m/sec from the speed of a boat which had been trapped in the tsunami (Kasahara and Chino, 1961). The energy for this scouring, transportation, and deposition of bottom materials is expected to be much greater than the energy dissipation due to bottom friction.

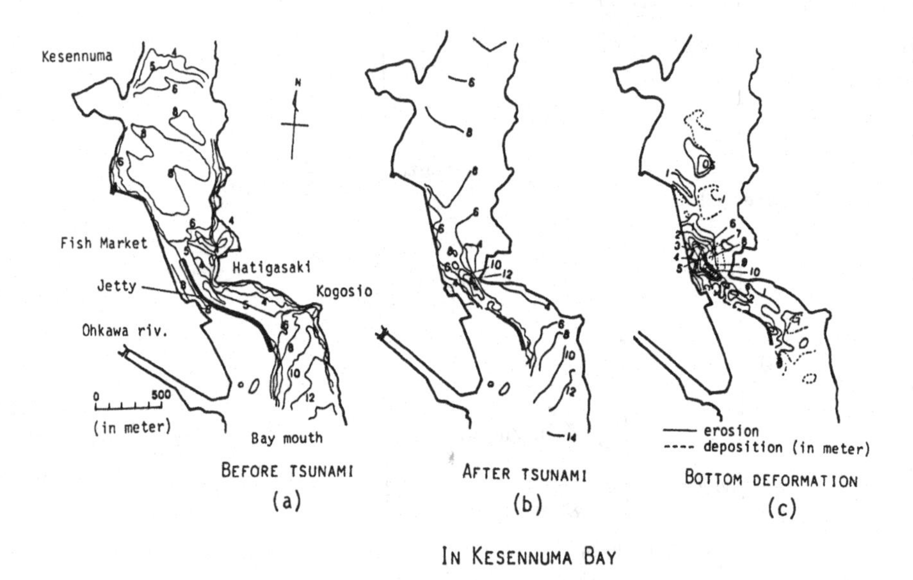

Fig. 10. Sea bottom contours in Kesennuma Bay, (a) before and (b) after the Chilean tsunami. (c) Change of water depth in meters.

5. Damage Done to Pearl Culture Rafts

5.1. *Accuracy of Velocity Computation*

Pearl culture was heavily damaged by the Chilean tsunami. Data on damage were collected in Matoya Bay, Kii Peninsula, in the middle of Japan (Figure 11). Because of its complicated topography as a ria coast, the current system in Matoya Bay is not simple.

Different from the water level, Aida's (1977) criterion in terms of K and κ are not applicable to the judgement of accuracy of the simulated current velocity, because current velocity has never been measured during tsunamis. We, therefore, apply the Richardson method that the deviation from the true value decreases monotonously

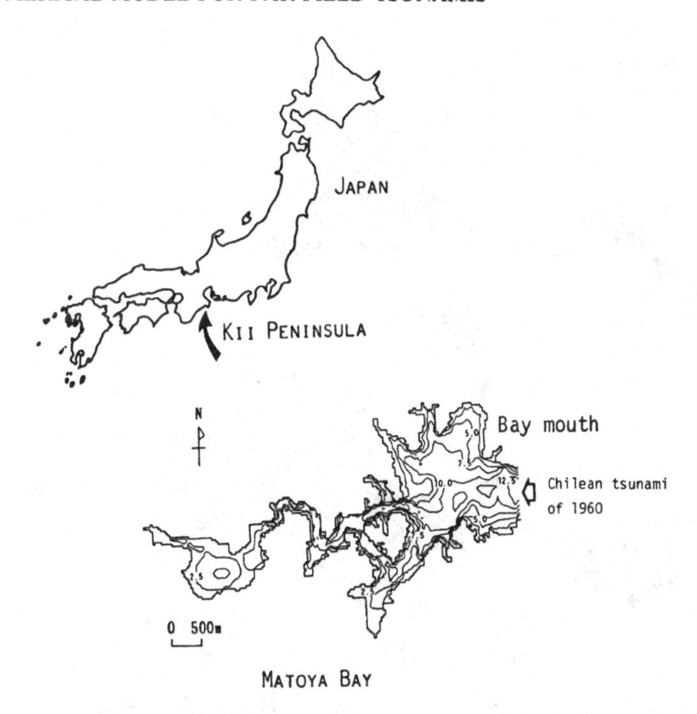

Fig. 11. Location and topography of Matoya Bay.

as the grid length, which plays a most important part in the numerical results, decreases. A power law is assumed to the computed maximum value at any point

$$u - u_1 = A(\Delta x_1)^p,$$
$$u - u_2 = A(\Delta x_2)^p, \qquad (5)$$
$$u - u_3 = A(\Delta x_3)^p,$$

in which u is the 'true' value, A and p are coefficients, u_1, u_2 and u_3 the computed results for $\Delta x_1 = 200$ m, $\Delta x_2 = 100$ m, and $\Delta x_3 = 50$ m. After the 'true' value is determined, the deviation of the computed values from it is computed and shown in Figures 12 and 13 for water level and current velocity. As the grid length is made
in which u is the 'true' value, A and p are coefficients, u_1, u_2 and u_3 the computed results for $\Delta x_1 = 200$ m, $\Delta x_2 = 100$ m, and $\Delta x_3 = 50$ m. After the 'true' value is determined, the deviation of the computed values from it is computed and shown in Figures 12 and 13 for water level and current velocity. As the grid length is made
finer, the maximum water level approaches the 'true' value. The current velocity shows a similar tendency as a whole. However, exceptions are found in narrow channels and the points where flows from branch bays meet.

Generally speaking, the current computation requires finer grids than the water-level computation. In order to obtain results of less than 10% error, the grid length

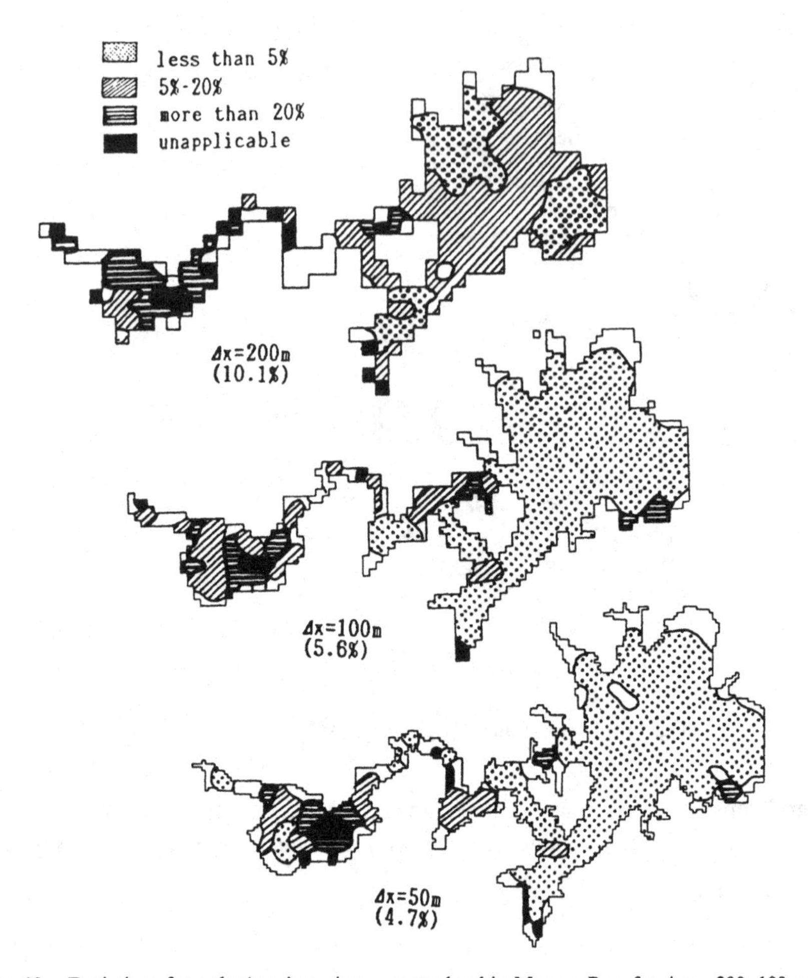

Fig. 12. Deviations from the 'true' maximum water level in Matoya Bay, for $\Delta x = 200$, 100, and 50 m.

is required to be less than 50 m. This is smaller than the grid length (about 200 m) required for the water level for the same order of accuracy.

5.2. *Damages Done to Pearl Culture Rafts*

Figure 14 gives the distributions of the computed maximum velocity and the degree of damage done to pearl culture rafts which was investigated after the Chilean tsunami (Sato, 1960). The expression 'damaged' means that a raft was washed away from its original position and destroyed or sunk. Then, more than 70% of the mother shells were lost. When a raft was partially damaged, i.e. if it is washed away and collides with other rafts, 20–30% of the mother shells were lost from the raft. When a raft was only moved but not washed away, it was not damaged and the

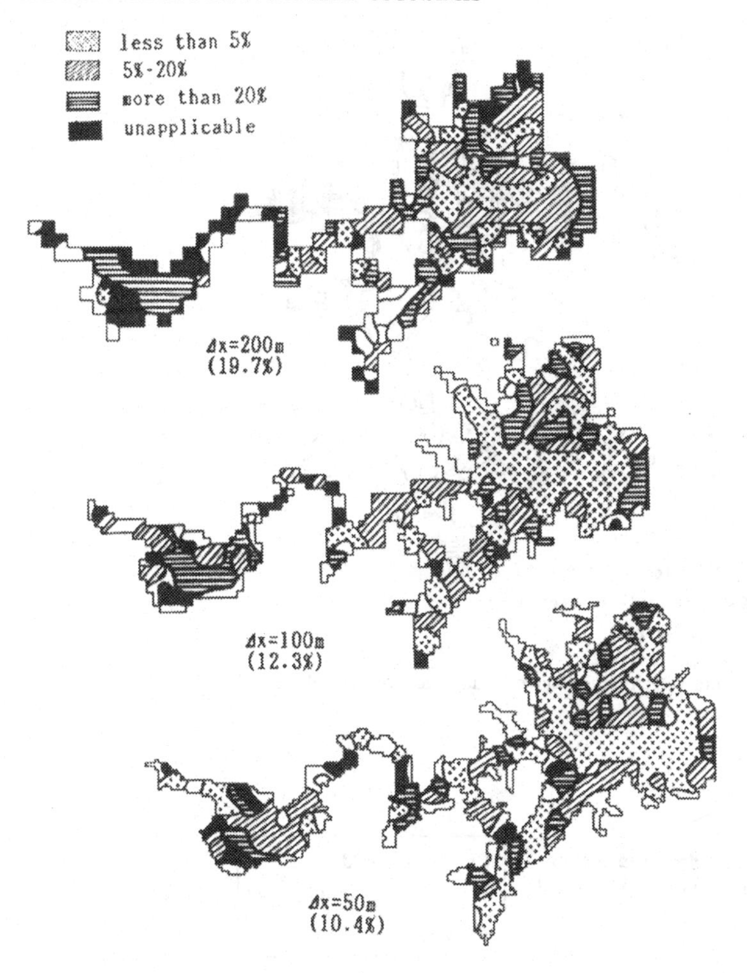

Fig. 13. Deviations from the 'true' maximum current velocity in Matoya Bay for $\Delta x = 200$, 100, and 50 m.

mother shells were safe. Recovery percentage of mother shells was 40 to 60%. Once they had fallen onto the silty sea bottom, the mother shells were thickly covered by silt which was violently moved and transported by tsunami currents. The fallen mother shells should be recovered within 2 or 3 days, otherwise, all of them will be found dead. If it was a stony bottom, the recovery percentage was higher (Sato, 1960).

A criterion of damage is given in Figure 15. Circles are without damage, triangles are partially damaged or moved, and crosses are completely destroyed or washed away. The maximum water level has no effect on the degree of damage, because loose mooring of rafts could follow a change in water level. If the maximum velocity does not exceed 1 m/sec, pearl culture rafts are safe. This is an ironic

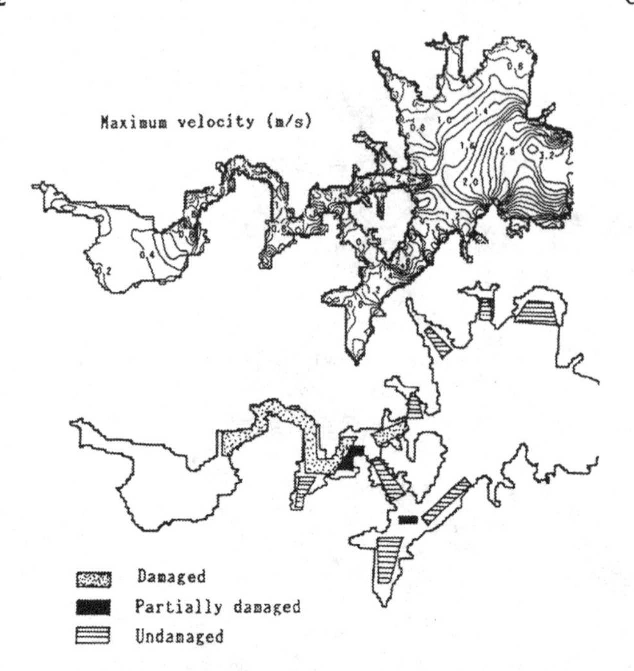

Fig. 14. Maximum velocity and degree of damage to aquaculture rafts.

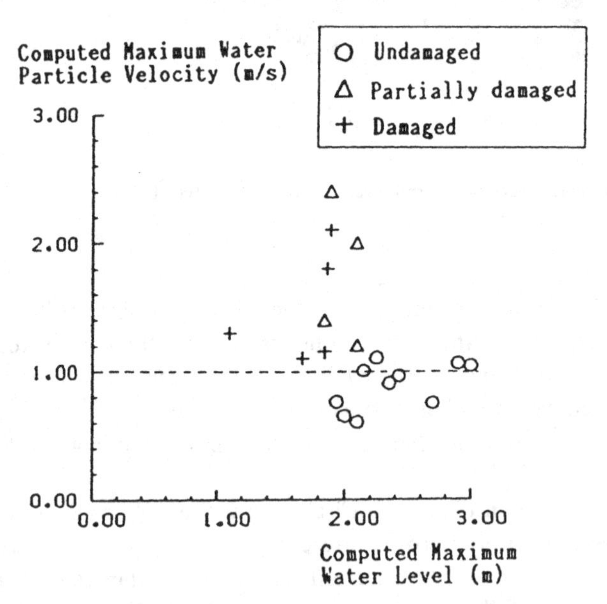

Fig. 15. A criterion of no damage done to pearl culture rafts in terms of the maximum water level and current velocity.

conclusion. A place where the tsunami current did not exceed 1 m/sec is not a good culture field because of poor sea-water exchange.

6. Conclusions

The Chilean tsunami of 1960 has been successfully simulated. In the ocean propagation computation, the linear longwave equation gives results of the same accuracy as the linear Boussinesq equation if the Imamura number (Equation (1)) is set nearly equal to unity. The CPU time is very much reduced by this method.

The overall distribution of the tsunami height along the Japanese Archipelago is simulated fairly well, except for near Hokkaido where the computed tsunami height is slightly higher than the measured height. Further analysis is required for the initial profile as well as the local topographical effect.

The coast transformation computation gives good agreement between the computed and measured results along the Sanriku coast, except for a few long bays. Disagreement may be caused by the insufficient evaluation of energy dissipation mechanisms such as severe scouring of sea bottom due to tsunami current.

A far-field tsunami has a longer period than a near-field tsunami. Amplification mechanism in the near-shore zone is therefore different. For the former, resonance in a bay is the most dominant. For the latter, shoaling and concentration are more important.

A criterion has been obtained for the damage done to aquaculture rafts. The computed maximum current velocity of 1 m/sec is the boundary between undamaged and damaged.

Appendix

The linearized and vertically integrated equations for long waves in one-dimensional problem is given as follows

$$\frac{\partial \eta}{\partial t} + \frac{\partial M}{\partial x} = 0, \qquad \frac{\partial M}{\partial t} + C_0^2 \frac{\partial \eta}{\partial x} = 0, \tag{A1}$$

where η is the water surface elevation, M the water discharge, $C_0 = \sqrt{gh}$ the celerity of the linear long wave, g the acceleration of gravity, and h the water depth. If the Staggered Leapfrog scheme, which is widely used for the simulations of long waves, is applied to Equation (A1), the discretized equations are obtained as

$$\frac{1}{\Delta t}[\eta_{j+1/2}^{n+1/2} - \eta_{j+1/2}^{n-1/2}] + \frac{1}{\Delta x}[M_{j+1}^{n} - M_j^n] = 0,$$

$$\frac{1}{\Delta t}[M_j^{n+1} - M_j^n] + \frac{C_0^2}{\Delta x}[\eta_{j+1/2}^{n+1/2} - \eta_{j-1/2}^{n+1/2}] = 0. \qquad \text{A2)}$$

where, Δt and Δx are grid lengths in the directions of t and x. The point (t, x) is replaced by $(n\,\Delta t, j\,\Delta x)$.

By using the Taylor series expansion, the discretized equations can be approximated by the modified differential equations. Equation (A2) is rewritten by the Taylor series at the point $(n\,\Delta t, (j+1/2)\,\Delta x)$ as follows (Imamura and Goto (1988)],

$$\frac{\partial \eta}{\partial t} + \frac{\partial M}{\partial x} + \sum_{m=1}^{\infty} \frac{1}{(2m+1)!}\left[\left(\frac{\Delta t}{2}\right)^{2m} \frac{\partial^{2m+1}\eta}{\partial t^{2m+1}} + \left(\frac{\Delta x}{2}\right)^{2m} \frac{\partial^{2m+1} M}{\partial x^{2m+1}}\right] = 0,$$

$$\frac{\partial M}{\partial t} + C_0^2 \frac{\partial \eta}{\partial x} + \sum_{m=1}^{\infty} \frac{1}{(2m+1)!}\left[\left(\frac{\Delta t}{2}\right)^{2m} \frac{\partial^{2m+1} M}{\partial t^{2m+1}} + C_0^2 \left(\frac{\Delta x}{2}\right)^{2m} \frac{\partial^{2m+1}\eta}{\partial x^{2m+1}}\right] = 0. \tag{A3}$$

It is found that the third term in Equation (A3) results from the discretization and is added to the original equations of the linear longwave theory. Elimination of M in equation (A3) yields the following differential equation by using the relationship for the linear longwave theory

$$\frac{\partial^2 \eta}{\partial t^2} - C_0^2 \frac{\partial^2 \eta}{\partial x^2} - C_0^2 \sum_{m=1}^{\infty} \frac{2}{(2m+1)!}\left(\frac{\Delta x}{2}\right)^{2m} (1 - K^{2m}) \frac{\partial^{2m+2}\eta}{\partial x^{2m+2}} = 0 \tag{A4}$$

where K is the Courant number defined as $K = C_0\,\Delta t/\Delta x$.

The last term in Equation (A4) consists of even derivaties only which have the dispersive effect that the wave propagation velocity decreases as the wave number increases. If only the major component of the last term in Equation (A4) is used, we obtain the following Equation (A5) as an approximation.

$$\frac{\partial^2 \eta}{\partial t^2} - C_0^2 \frac{\partial^2 \eta}{\partial x^2} - \frac{C_0^2\,\Delta x^2}{12}(1 - K^2)\frac{\partial^4 \eta}{\partial x^4} = 0 \tag{A5}$$

Figure 16 compared wave celerity characteristics of Equations (A4) and (A5), by dotted and broken lines. The ordinate is nondimensionalized by the physical celerity of the linear long wave. The abscissa is the wave number nondimensionalized by the spatial grid length. Since this figure confirms that Equation (A5) is a good approximation to Equation (A4), Equation (A5) is used in the following discussion.

On the other hand, the linear Boussinesq equation including the physically correct dispersion effect is given as follows.

$$\frac{\partial^2 \eta}{\partial t^2} - C_0^2 \frac{\partial^2 \eta}{\partial x^2} - \frac{C_0^2 h^2}{3}\frac{\partial^4 \eta}{\partial x^4} = 0 \tag{A6}$$

The last term is the dispersion term of the first-order approximation. Comparing Equations (A5), (A6), it is immediately understood that if the coefficients of the two equations are of the same value, or if the value given by Equation (A7) is taken to be equal to unity, the numerical results of the linear longwave equation give the solution of the linear Boussinesq equation.

$$\mathrm{Im} = \sqrt{\frac{\text{Numerical term}}{\text{Physical term}}} = \frac{\Delta x}{2h}\sqrt{1 - K^2}. \tag{A7}$$

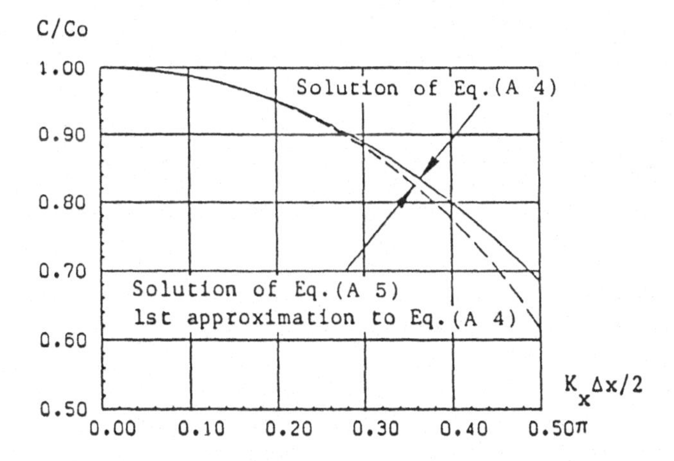

Fig. 16. Numerical dispersion effect for the staggered leapfrog scheme with $K = 0.5$. C: numerical wave celerity, $K_x\ \Delta x/2$ = nondimensional wave number.

Acknowledgement

A part of the reported research was supported by grant No. 01601005 from the Ministry of Education, Science and Culture, Japan.

References

Aida, I.: 1977, Reliability of a tsunami source model derived from fault parameters. *J. Phys. Earth* **26**, 57–73.

Imamura, F., Goto, C., and Shuto, N.: 1987, Numerical simulation of the 1964 Alaskan tsunami including the dispersion term, *Proc. Internat. Tsunami Sympos.* NOAA, pp. 144–146.

Imamura, F. and Goto, C.: 1988, Truncation error in numerical tsunami simulation by the finite difference method, *Coastal Engng. in Japan, JSCE* **31**, 245–263.

Kajiura, K.: 1963, Effects of a breakwater on the oscillations of bay water, *Bull Earthq. Res. Inst.* **41**, 535–571 (in Japanese).

Kajiura, K.: 1963, The leading wave of a tsunami, *Bull. Earthq. Res. Inst.* **41**, 535–571.

Kanamori, H. and Ciper, J. J.: 1974, Focal process of the Great Chilean earthquake May 22, *Phys. Earth Planet. Inter.* **9**, 128–136.

Kasahara, K. and Chino, I.: 1961, Report of investigation, *Rep. on the Chilean Tsunami of May, 24, 1960, as Observed along the Coast of Japan,* Committee for the Field Investigation of the Chilean Tsunami of 1960, 280 (in Japanese).

Mansinha, L. and Smylie, D. E.: 1971, The displacement fields of inclined faults, *Bull. Seism. Soc. Amer.* **61**, 1433–1440.

Sato, T.: 1960, The Chilean Tsunami and damages to aquaculture rafts in Mie Prefecture, *Suisanzoshoku* **8**, 193–202 (in Japanese).

Shuto, J., Suzuki, T., Hasegawa, K. and Inagaki, K.: 1986, A study of numerical techniques on the tsunami propagation and runup., *Science Tsunami Hazard* **4**(2), 11–24.

Natural Hazards **4**: 257–266, 1991.

Tsunami Ascending in Rivers as an Undular Bore

YOSHINOBU TSUJI, TAKASHI YANUMA, ISAO MURATA,
and CHIZURU FUJIWARA
Earthquake Research Institute, University of Tokyo, Yayoi 1-1-1, Bunkyo-Ku, Tokyo 113, Japan

(Received: 15 March 1990; revised: 23 July 1990)

Abstract. At time of the 1983 Japan Sea tsunami, waves in the form of a bore ascended many rivers. In some cases, bores had the form of one initial wave with a train of smaller waves, and in other cases, such a wave train did not appear and only a step with a flat water surface behind was observed. In the present study, it is clarified that both undular-type and nonundular-type bores can be recognized as solutions of the KdV–Burger's equation which was introduced by Johnson in 1972. Numerically obtained analytical solutions and results of laboratory experiments are compared.

Key words. Bore, undular bore, hydraulic jump, KdV–Burgers equation, nonlinear and dispersive wave with energy dissipation, riverwall planning, breaking waves, phase plane method, Runge–Kutta–Gill's method, amplification of tsunami soliton dispersion.

1. Introduction

At the time of the Japan Sea earthquake tsunami of 1983, it was observed that tsunami waves ascended several rivers in the form of a bore. It was observed in some rivers that bores had a breaking front followed by a flat water surface, and in other rivers, undular bores occurred with an initial wave following by an undulated wave train (Figure 1). Chester (1966) called the former type of bore a 'strong bore' and the latter a 'weak bore' (Figure 2).

It is well known that the dynamics of a bore are essentially the same as those of a hydraulic jump. It is also known that when the ratio of the depth upstream to that downstream is bigger by about two, a strong bore is likely to appear and in the case where the ratio is close to unity, then a weak bore appears. In the case of a strong bore, the energy of the wave is strongly dissipated at the front step with a breaking crest; and sometimes a horizontal vortex can be observed.

Chester (1966) assumed that water is viscid and the velocity profile basically takes a parabolic distribution as Poiseulle's flow at any instance. He showed that the water surface is expressed approximately in the style of a hyperbolic tangential equation in both the up- and downstream sides and, under some conditions, the bore itself takes the form of Airy's function.

Freeman and Johnson (1970) dealt with the problem of a shallow water wave on currents with shear flow and deduced a modified Korteweg-deVries equation for the problem.

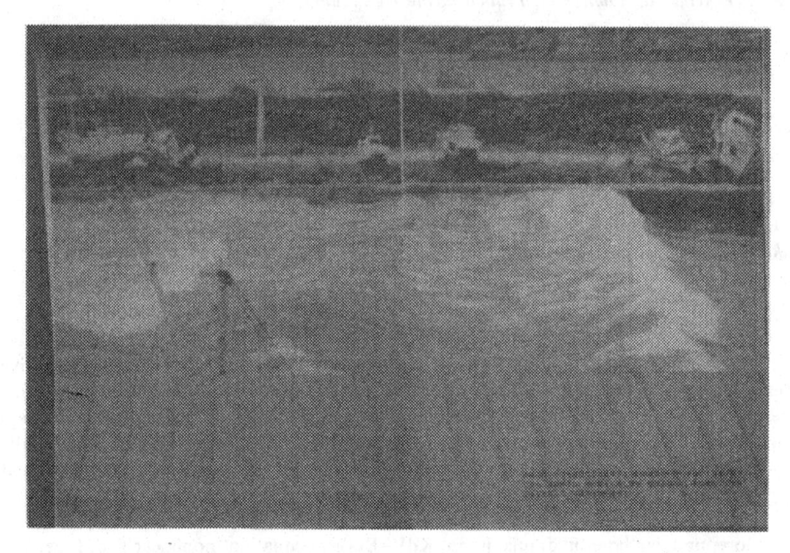

Fig. 1. An undular bore ascended along the small channel in Noshiro Port, Akita Prefecture, during the time of the 1983 Japan Sea tsunami. (By courtesy of the Hoku-U Shimpo-Sha Press.)

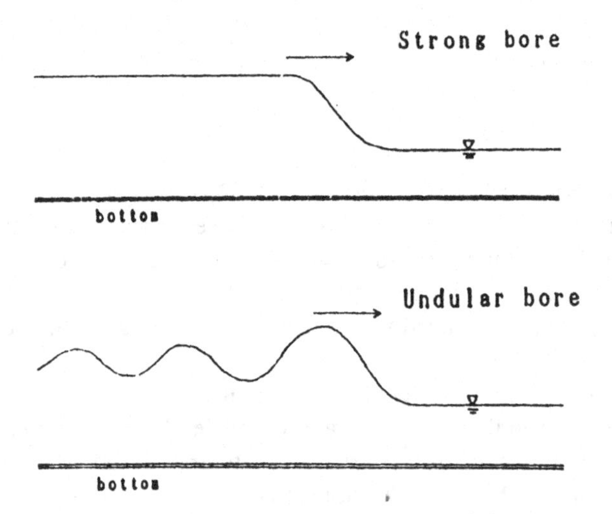

Fig. 2. Strong bore and undular (weak) bore.

Johnson (1970, 1972) assumed viscous fluid and showed that the surface profile of the bore is expressed by the KdV–Burgers equation as

$$\eta_t + \eta\eta_x + \beta\eta_{xxx} = p\eta_{xx}, \tag{1}$$

where η is the water surface height, p is a parameter of internal or bottom friction, and β is a constant relating to dispersion. The solution of Equation (1) would be asymptotically reduced to solitary or cnoidal waves in the case where viscosity vanished.

Pelinovskii (1982) discussed the KdV–Burgers equation in a more general style in his textbook of *Nonlinear Dynamics of Tsunami Waves*. He proposed the controlling equation as

$$\eta_t + c_0(1 + 3\eta/2D)\eta_x + c_0 D/6\, \eta_{xxx} = -F(\eta), \qquad (2)$$

where η is displacement of water surface, c_0 is longwave velocity $(=\sqrt{gD})$, D is depth, and $F(\eta)$ is a kind of friction function.

Generally speaking, if during tsunami wave propagation the wave energy dissipates, and nonlinear effects also take place, the control equation for the surface elevation would be in the form of (1) or (2). The causes of energy dissipation are considered not only by the internal molecular or *eddy* viscosity, but also by bottom friction at the river bed. The effectiveness of energy dissipation by friction and/or eddy viscosity are generally greater than that of molecular viscosity by several orders, and the coefficients of the energy decay term, that is, the left-hand sides of (1) and (2), are difficult to estimate exactly.

At the time that laboratory experiments of bores were conducted, it was sometimes observed that the leading crest was dispersed into two or more crests, which looks like the dispersion of solitons from a ridge under the system of the Korteweg–de Vries equation. For the dispersion of bore crests, which is controlled by the KdV–Burgers equation, the energy is continuously dissipated in contrast to the total energy which is conserved during the dispersion of solitons. The mean water-level changes in the front and rear parts of the bore, while that in the soliton system does not change. Therefore, we should strictly distinguish the crest dispersion of a bore from the fission of solitons.

Shokin *et al.* (1989) gave an example of a numerical result of wave dispersion in a nonlinear dispersive system.

In the present study, we assume the water surface satisfies the KdV–Burgers equation (1) and consider the method to obtain a numerically strict solution of it. A laboratory experiment of bores was conducted in a small water channel and the observed form of the bore was compared with a numerically calculated one.

2. Method of Calculation of Numerical Solution of the KdV–Burgers Equation

In the same way as the dispersion of the KdV equation, we assume that the ratio of wave height 'a' to depth D is a small value of the order of ε^1, and that the square of the ratio of depth D to the horizontal scale (effective wavelength) L is of the same order. In the case where viscosity or bottom friction can be neglected, we finally have the Korteweg–de Vries equation as

$$\eta_t + c_0\eta_x + 3c_0/2D\eta\eta_x + c_0 D^2/6\eta_{xxx} = 0, \qquad (3)$$

where η is water surface displacement, x is the horizontal co-ordinate, t is time, c_0 is the velocity of long wave $(=\sqrt{gD})$, and g is the acceleration of gravity. If we do not neglect the effect of the dissipation of energy caused by internal viscosity and/or bottom friction, we have the KdV–Burgers equation as

$$\eta_t + c_0\eta_x + 3c_0/2D\eta\eta_x + c_0D^2/6\eta_{xxx} = R\eta_{xx}, \tag{4}$$

where R is the coefficient of energy dissipation. In the case where the energy dissipation is caused by molecular viscosity and velocities of water particles at the bottom are zero in both vertical and horizontal directions, R is given in the form

$$R = Q\nu L^2/D^2 \tag{5}$$

where ν is coefficient of dynamic viscosity, L is the wavelength, and Q is a constant. Chester showed that in the case of Poiseulle flow in both up- and downstream sides, Q is expressed as

$$Q = (1 - 2/5F^2)/3, \tag{6}$$

where F is a kind of Froude number given by $F = U/\sqrt{gD}$, and U is the velocity of the bore.

By using transformations of variables as

$$\xi = (3c_0/2D)\eta, \qquad z = \beta(x - c_0t), \qquad \tau = \beta^{-1/2}t,$$

$$\delta = \beta^{-1/2}R \quad \text{and} \quad \beta = c_0D^2/6,$$

Equation (4) turns out to be

$$\xi_\tau + \xi\xi_z + \xi_{zzz} = \delta\xi_{zz}. \tag{7}$$

We assume that at the left-hand side (upstream direction) infinity ($z = -\infty$) ξ approaches asymptotically to zero, and at the right-hand side (sea-side direction, $z = +\infty$) infinitly to ξ_∞. We make one more transformation as

$$T = (z - \xi_\infty\tau/2)/\sqrt{8/\xi_\infty}, \qquad V = \xi/\xi_\infty, \tag{8}$$

and

$$p = 2\,\delta\sqrt{2/\xi_\infty}. \tag{9}$$

Equation (7) is rewritten into the style of an ordinary second-order differential equation as

$$V_{TT} + 4V(V-1) = pV, \tag{10}$$

where p is parameter proportional to the viscosity.

We cannot obtain the analytical solution of Equation (10), but we can obtain a numerical solution by using the phase-plane method (Figure 3). We put $W = \mathrm{d}V/\mathrm{d}T$, and we have the relationship between V and W as

$$\mathrm{d}W/\mathrm{d}V = \{pW - 4V(V-1)\}/W. \tag{11}$$

There are two singular points on the phase plane: $(V, W) = (0, 0)$ and $(1, 0)$. Johnson (1970) showed that if p satisfies the condition $0 \leqslant p \leqslant 4$, one characteristic

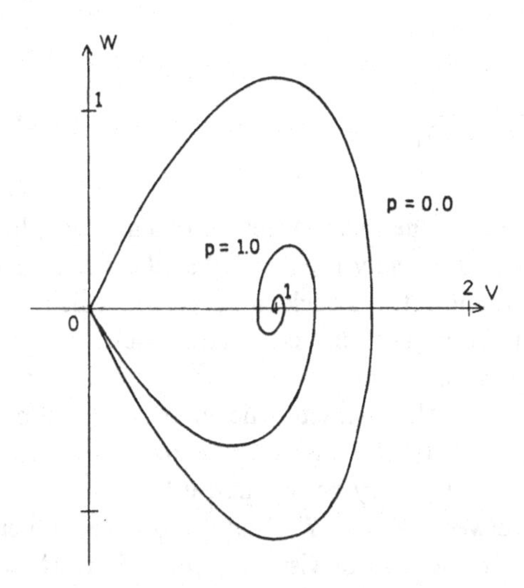

Fig. 3. Phase plane (V, W).

curve would be decided for each value of p, and that the characteristic curve starts from the original point (0, 0) and is terminated at the other singular point (1, 0) except when the characteristic curve corresponds to $p = 0$.

If we give some value of p, we can numerically calculate the exact shape of the characteristic curve. We take the angle of the radiation direction of a characteristic curve from the origin (0, 0) as θ. If we select a point P on the curve as being close to the origin, the ordinates (V, W) have the relation

$$\mathrm{d}W/\mathrm{d}V = W/V = \tan\theta. \tag{12}$$

At a point close to the origin, V-1 in (11) approaches to -1, and so we have

$$\tan\theta = p + 4\cot\theta, \tag{13}$$

which is a quadratic equation of $\tan\theta$ and easily solved as

$$\tan\theta = \{\pm\sqrt{(p^2+16)} - p\}/2. \tag{14}$$

We can decide the radiation direction of the characteristic curve for a given p. Whole parts of the characteristic curve can be obtained by Runge–Kutta–Gill's Method by alternatively using of (11) and its inverse relation

$$\mathrm{d}V/\mathrm{d}W = W/\{pW - 4V(V-1)\} \tag{15}$$

avoiding a denominator close to zero.

The function that expresses the characteristic curve (V, W) is multivalued, but if we put the relation in an explicit style of a function as $W = G(V)$, we can obtain V as a function of T by connecting the following two integrated values

$$T_1 = \int \frac{\mathrm{d}V}{G(V)}, \tag{16a}$$

and

$$T_2 = \int \frac{\mathrm{d}W}{4pW - 4G^{-1}(W)\{G^{-1}(W) - 1)\}}, \tag{16b}$$

alternatively.

For the case of $p = 0$, that is, the case of the inviscid fluid problem, the solution reduces to a soliton exactly, and it is theoretically predicted that the characteristic curve for $p = 0$ passes the point (1.5, 0) and returns back to the origin. This fact can be used for checking the accuracy of the numerically calculated result.

In the present study, the interval step for V or W on the numerical integral was selected as 0.001, and it was confirmed that the characteristic curve for $p = 0$ passes (1.4997, 0.0) and returns back to (0.0005, 0.0). Thus, we can recognize that the numerical calculation was made with an accuracy better than 0.1%.

Figure 4 shows the relationship between V and T obtained by the numerical calculation. It should be noticed that in the case of viscosity, p is relatively small, the bore takes the undular type weak bore and, in contrast, where p becomes

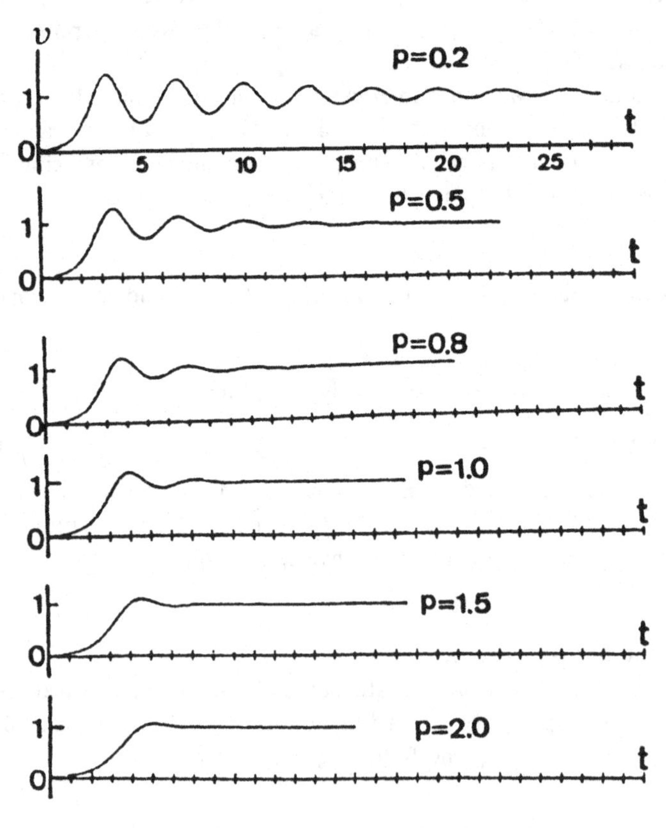

Fig. 4. Numerical solutions of Equation (10).

greater, it becomes similar to the shape of a strong bore. We can actually recognize that in the case of $p > 1$, the bore is of the strong type.

We put the function shown in Figure 4 as $V = F_p(T)$. The dimensional expression of the bore shape is given by

$$\eta = \eta_\infty F_p(T), \tag{17}$$

where

$$T = [x - c_0\{1 + (3\eta_\infty/4D)t\}]/\sqrt{8D^2/9\eta_\infty}. \tag{18}$$

We also notice that in the case of undular bores (p: small), the height of the initial crest becomes higher than that of the right side infinitive, that is, the height of the upheaved sea surface caused by the tsunami. If p is close to zero, then the height would be converged to 1.5 times the height of the tsunami.

Figure 5 shows the amplified ratio of the initial crest to the energy dissipation parameter p. We should notice that in the soliton dispersion problem in the KdV system, the height of the initial soliton could be close to twice the height of the original wave, and that the KdV–Burgers system resembles the KdV system, but is essentially different.

We should notice that the water-surface shape (intervals of crests, heights of first, second crests, and so on) is decided completely only if the values of p are given. Values of p are decided by the relative difference of the depths of both sides and by viscosity or friction on the bed.

It is not permitted that the function of $F_p(T)$ would be enlarged arbitrarily both in horizontal and vertical directions.

3. Comparison with Laboratory Experiment

In order to check the above-mentioned theoretical discussion, an indoor experiment was conducted. Figure 6 shows the schematic drawing of the total apparatus in

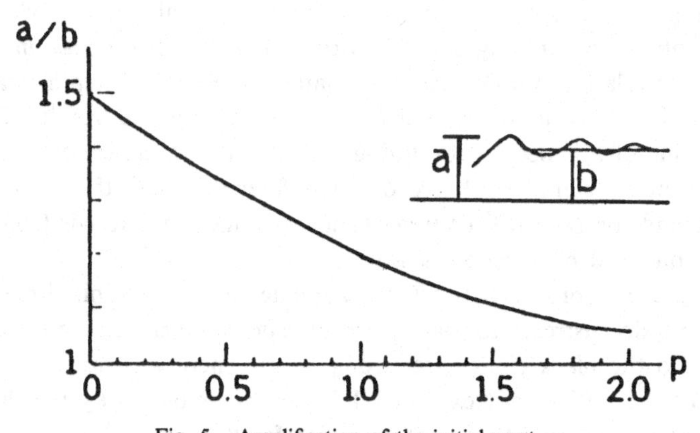

Fig. 5. Amplification of the initial crest.

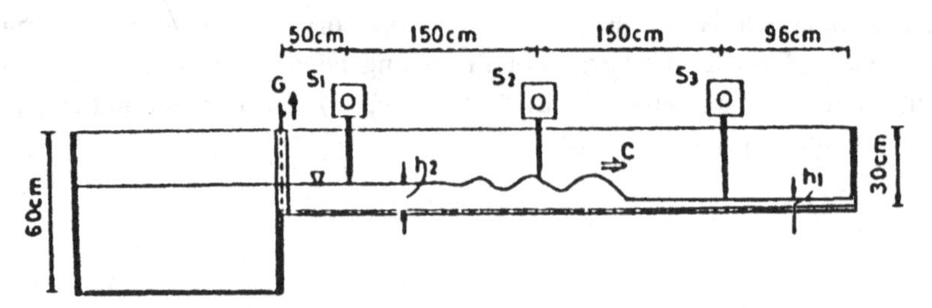

Fig. 6. Schematic drawing of the experimental apparatus.

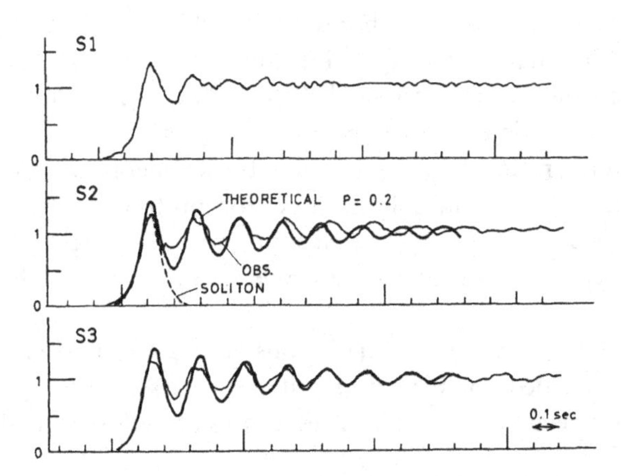

Fig. 7. Comparison of experimental and theoretical bores.

which there is a small channel with width of 7 cm and length of 4.46 cm. On one side of the channel, there is a rectangular pool with a size of 0.98 m^2 with a depth of 60 cm. We used three water level gauges of the surface-follower type. The accuracy of these gauges is 0.1 mm.

Figure 7 shows one example of the result. S1, S2, and S3 are the servo sensors shown in Figure 6. The thick lines in the center and lower figures show the observed water-surface changes, and the thin lines are the theoretical results. We can recognize that the observed curve agrees well with that of the theoretical one.

Figure 8 shows the relationship between the relative difference of depth, that is, tsunami height, and the value of p, which is evaluated by the result of the experiment. The value of p is decided by fitting the intervals of the crests to those of the theoretical one. When the relative depth difference is less than 0.6, p is smaller than unity and the bore will be weak (undular), and if it exceeds 0.6, then p become close to one and bore becomes strong.

Figure 9 shows the relationship between the amplification of the initial crest to the ratio of the depth downstream to that upstream. The white circles are for cases of undular bores and the black circles are strong bores. There was no case where the amplification of the initial crest exceeded 1.5, as was predicted by the theory and, moreover, the tsunami height was not amplified for the cases of strong bores.

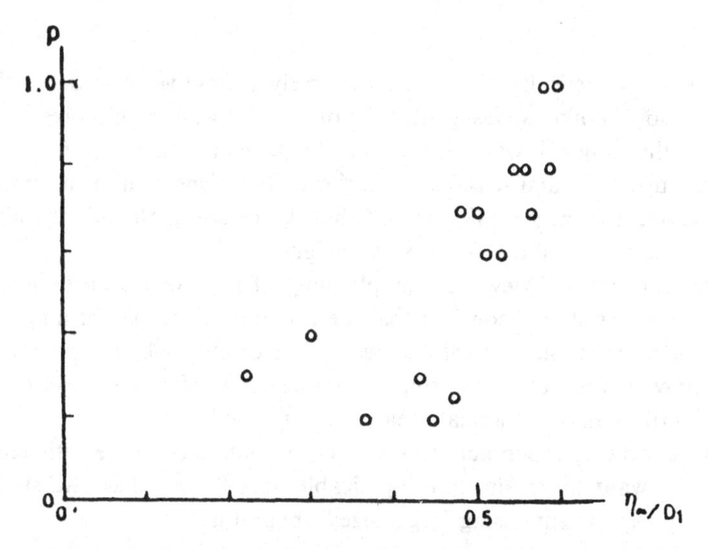

Fig. 8. Relationship between p in Equation (10) and the relative tsunami height.

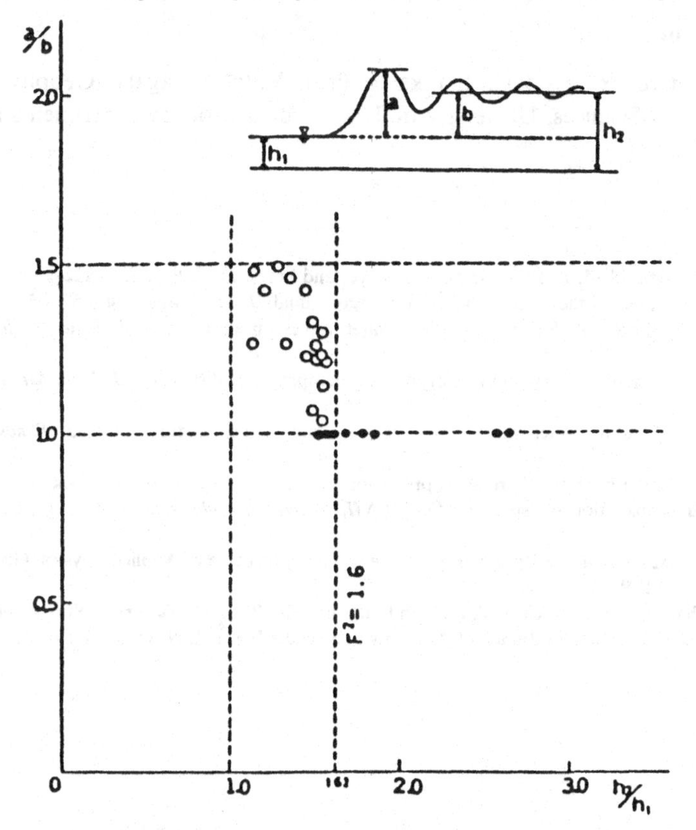

Fig. 9. Relationship between the amplification of the initial ridge and the relative tsunami height. White circles indicate undular bores; black circles indicate strong bores.

4. Discussion

We confirmed, both theoretically and under extremely restrictive conditions, that the height of the soliton-like waves generated by the dispersion of bores never exceed 1.5 times the original wave height of the tsunami, that is, the depth difference between upstream and seaside. As in the soliton dispersion in the system of the KdV equation, it is theoretically proved that the height of the initial soliton can be close to twice that of the original wave height.

From the practical point of view, for the planning of river or channel walls of both sides of a stream, we should consider that the tsunami height may be amplified up to 1.5 times the initial tsunami height in deep rivers or channels. If we plan the height of the river walls without considering the amplification effect for weak bores, sea water may overflow into residential areas at the riverside.

The size of the laboratory experiment was much too small to compare with actual bores, and so if we want to obtain more applicable data for p values, we should conduct laboratory experiments using larger-sized apparatus.

Acknowledgement

The authors wish to express their thanks to Prof. Yutaka Nagata, Geophysical Institute, Faculty of Sciences, University of Tokyo, for laboratory experiments and active discussions.

References

Benjamin, T. B. and Lighthill M. J.: 1954, On cnoidal waves and bores, *Proc. Roy. Soc.* **A224,** 448–460.

Chester, W.: 1966, A model of the undular bore on a viscous fluid, *J. Fluid Mech.* **24,** 367–377.

Freeman, N. C. and Johnson R. S.: 1970, Shallow water waves on shear flows, *J. Fluid Mech.* **42,** 401–409.

Johnson, R. S.: 1970, A nonlinear equation incorporating damping and dispersion, *J. Fluid Mech.* **42,** 49–60.

Johnson R. S.: 1972, Shallow water waves on a viscous fluid – The undular bore, *J. Fluids* **15,** 1693–1699.

Nakamura S.: 1973, On the hydraulic bore and application of the results of its studies to the problems of generation and propagation of tsunamis, *Sakh-KNII, Yuzhno-Sakhalinsk,* **32,** 129–151 (in Russian).

Pelinovskii, Ye, N.: 1982, *Non-linear Dynamics of Tsunami Waves,* Institute of Applied Physics, Gor'kii, Academia Nauk, U.S.S.R, (in Russian).

Shokin Yu. I., Chubarov, L. B., Marchuk, An. G. and Simonov, K. B.: 1989, *Numerical Experiments in the Problems of a Tsunami,* Siberia Branch of Academy of Sciences U.S.S.R, Novosibirsk (in Russian).

Natural Hazards **4**: 267–283, 1991.

Assessment of Tsunami Hazard in the Italian Seas

STEFANO TINTI
Dipartimento di Fisica, Settore di Geofisica, Università di Bologna, Viale Berti Pichat 8, 40127 Bologna, Italy

(Received: 18 June 1990; revised: 19 July 1990)

Abstract. A method for the evaluation of tsunami potential in the seas surrounding Italy is presented. A major difficulty for performing reliable estimates of tsunami occurrence is that the existing tsunami catalog for Italy includes a small number of cases. This is due partly to the catalog incompleteness, strangely more pronounced in our century, and partly to the relative infrequency of tsunamis along the Italian seas. Evaluation of tsunami activity is therefore deduced by complementing the tsunami catalog data with data on seismicity that are by far more abundant and reliable. Analysis of seismicity and assessment of earthquake rate in coastal and submarine regions form the basis of the present method to perform tsunami potential estimates for Italy. One essential limitation of the method is that only tsunamis of seismic origin are taken into account, which leads to an underestimation of the tsunami potential. Since tsunamis generated by earthquakes are much more frequent than events produced by slumps or volcanic eruptions, the underestimation is not dramatic and very likely affects only a limited portion of the Italian coasts. In the present application of the method, eight separate regions have been considered that together cover all the coasts of Italy. In each region, seismicity has been independently examined and the earthquake potential has been calculated in small 20′ × 20′ cells. Then, on the basis of suitable assumptions, tsunami potential has been evaluated in each cell. According to this study, the Italian coasts that are the most exposed to the attacks of locally generated tsunamis are to be found in the Messina Straits, in Tyrrhenian coasts of Calabria, in the Ionian Sicilian coasts around Catania, and in the Gargano promontory in the Southern Adriatic Sea. Furthermore, this study confirms that the Northern Adriatic Sea has a low level of tsunami potential, in agreement with recent studies emphasizing that the large historical events concerning this region included in the first versions of the Italian tsunami catalog are largely overestimated and must be decreased.

Key words. Tsunamis, earthquakes, hazard assessment, Italian Seas.

1. Introduction

Occurrence of tsunamis in the Mediterranean Sea is known since the ancient times of Greek and Roman civilizations. The Santorin explosion at the epoch of the Minoan supremacy in the Aegean Sea (about 1500–1400 BC) and the 21 July 365 submarine Cretan earthquake are the most remarkable examples of events producing tsunamis that invested the whole Mediterranean basin. According to the existing tsunami catalogs, it is known that in historical times large tsunamis were generated in various parts of the Mediterranean, and mostly in the Central and Eastern domains. This consequently shows that evaluation of tsunami hazard is a necessary step for any studies concerning the hazards in the Mediterranean that derive from natural sources.

The present investigation addresses the problem of evaluating tsunami potential in the seas surrounding Italy and, to the author's knowledge, constitutes the first systematic approach on this subject. The determination of the tsunamigenic sources and the evaluation of their potential is a classic task that has important practical implications on coastal economy defence and planning as well as on population protection. A traditional tool to achieve this task is the statistical analysis of the tsunami sequences reported in the catalogs. This method has been widely applied in those countries, such as Japan, where tsunami activity is particularly high. The reliability of statistical studies depends on the quality and the quantity of the available data. The Mediterranean tsunami dataset is insufficient to allow the indiscriminate application of statistical methods. For example, the total number of events reported in the first version of the Italian tsunami catalog is 154 (Caputo and Faita, 1984), increased to 170 in the subsequent updated version (Bedosti and Caputo, 1986). If one considers that as many as 51 reports should be taken as extremely dubious cases, the number of certain events reduces to 119, that may be classed according to the tsunami origin as follows: 92 produced by earthquakes, 25 by volcanic eruptions, and 2 by slumps. Since the majority of the tsunamis were generated by local earthquakes and since the seismic datasets are generally fairly large and often of better quality than tsunami catalogs, it is clear that the evaluation of coastal and submarine seismicity is extremely advantageous for the ultimate evaluation of the tsunami potential.

The spatial distribution of the tsunamis generated around Italy is shown in Figure 1, derived mainly on the basis of the above-mentioned Italian tsunami catalogs with some minor corrections introduced by the author. However, the results of the recent revisions, especially affecting the events in the Northern Adriatic (see Tinti and Guidoboni, 1988; Guidoboni and Tinti, 1989), are not included. As may be seen, the reported tsunamigenic earthquakes, denoted by open circles, have epicenters mostly on the coast and sometimes on land far from the coastline, and occur in all seas around Italy. Volcanic tsunamis, denoted by open squares, are found in the Tyrrhenian and Southern regions and are predominantly related to the activity of Vesuvius, Etna, and Aeolian Islands volcanoes. Several anomalous sea perturbations of uncertain origin are reported in all seas, but especially in the Gulf of Naples and in the Ligurian Sea: very unlikely they are tsunamis since they cannot be associated to any clearly suitable exciting mechanisms. Looking at Figure 1, the most important observation that can be made is that the events may be easily grouped in clusters, defining separate sources. This enables us to perform a suitable partition of the Italian coasts into distinct regions. The selected partition into eight tsunami generation domains is shown in Figure 1 as well. Italian Adriatic coasts are subdivided into 3 regions, numbered 1 to 3 in the figure; the region 4 encompasses the Calabrian Arc tsunamigenic sources; the region 5 includes Western Sicily; and regions 6 to 8 cover the Tyrrhenian and the Ligurian coasts. The few events falling outside the partition either pertain to the Eastern Adriatic generation sources or are reported to be associated to epicenters too far from the sea; hence in the latter case, doubts may be reasonably raised on the reliability of the epicentral coordinates or even of the reports themselves. The

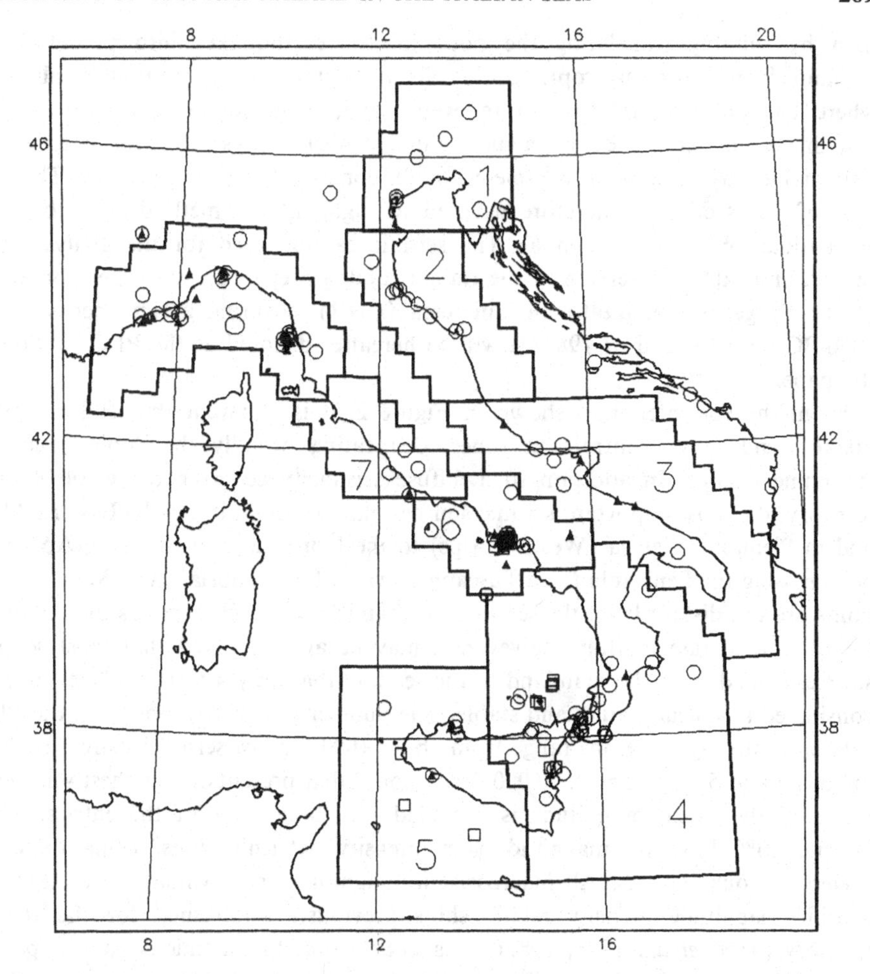

Fig. 1. Italian tsunami map. It is mostly based on the Italian tsunami catalogs with some corrections of the author. Open circles are tsunamigenic earthquake epicenters, open squares are tsunamigenic volcanic events, the few open triangles are tsunamis produced by landslides or slumps, while solid triangles are dubious tsunamis. In this investigation Italian coastal areas are partitioned into eight regions, delimited by thick solid-line borders in the figure; they divide the main tsunami sources from one another.

partition shown is the first step of the analysis carried out in the present study. Seismicity and then tsunami potential will be evaluated independently in each of the above regions by using a method illustrated in the next section. Results will be presented and discussed in the last section of the paper.

2. Illustration of the Method of Analysis

As pointed out in the previous section, estimates of seismic tsunami potential in each region will be determined by evaluating the seismic potential in the region and then by suitably translating the potential for earthquakes into potential for tsunamis by means of appropriate relations. It should be observed that for all areas where tsunamis are relatively rare events and, as a consequence, the known tsunamic events are to few for a successful statistical analysis, this seems to be the only viable way to perform reliable estimates concerning tsunami activity. The first part of this section is therefore devoted to highlight the method used to make evaluations of seismic potential. The seismic catalog used for the analysis is a revised and extended version of the Italian catalog prepared within the framework of the Progetto Finalizzato Geodinamica (CNR), covering a time period from 1450 BC until the end of 1982. It will be hereafter denoted as the PFG catalog in the paper.

From the epicenter map shown in Figure 2, it is apparent that Italian earthquakes tend to concentrate in zones, delineating broadly the Alpine and the Apennine Chains. In addition to a diffuse seismicity covering the whole Italian territory, the most important seismic sources may be found in Friuli (Eastern Alps) and in Piemonte-Liguria (Western Alps) to the North, and in the main Apennine belt crossing the Peninsula from Tuscany down to the Calabrian Arc. Most seismic faults are too distant from the sea to be tsunamigenic, which explains the relatively small tsunami rate in Italy; however, it may be also observed that most seismic sources extend from the mainland to the sea and that only a few coastlines may be considered seismically quiet and stable. The number of events reported in the PFG catalog in the eight regions range from about 1000 for low seismicity areas such as regions 3 and 5 up to about 10 000 for region 4 that presents the highest seismicity rate in Italy. Local magnitude is provided for about 20% of the entries, while around 10% have no magnitude nor intensity determinations, being therefore useless for our analysis. All macroseismic magnitudes are evaluated by means of specific magnitude-intensity relationships previously established for the Italian territory (Tinti *et al.*, 1986, 1987). It is known that the seismicity rate, apparent from seismic catalogs, departs from the true rate mainly due to catalog incompleteness; this affects different portions of the earthquake catalogs in a different way. In general, this effect is more severe for the first catalog sections comprising the historical earthquakes than for the most recent periods that include the instrumentally recorded quakes. Catalog completeness, however, is not only a function of time, but also of the earthquake size. In fact, the largest shocks, hardly escaping observation, are more likely included in the compilations. In order to analyze the completeness of the seismic series in each one of the eight regions, a method has been used that is applicable on sequences of main shocks and on magnitude classes. In the present application, magnitude partition depends on the region, but generally, the considered classes are 0.6 wide in magnitude unit and the smallest class ranges from 2.4 to 3. The method, which is a variant of the one devised by Tinti and Mulargia (1985), has the threefold final goal of evaluating the completeness period for each magnitude class, of estimating the true seismicity rate, and eventually of measuring the incompleteness level of any portion of the catalogue.

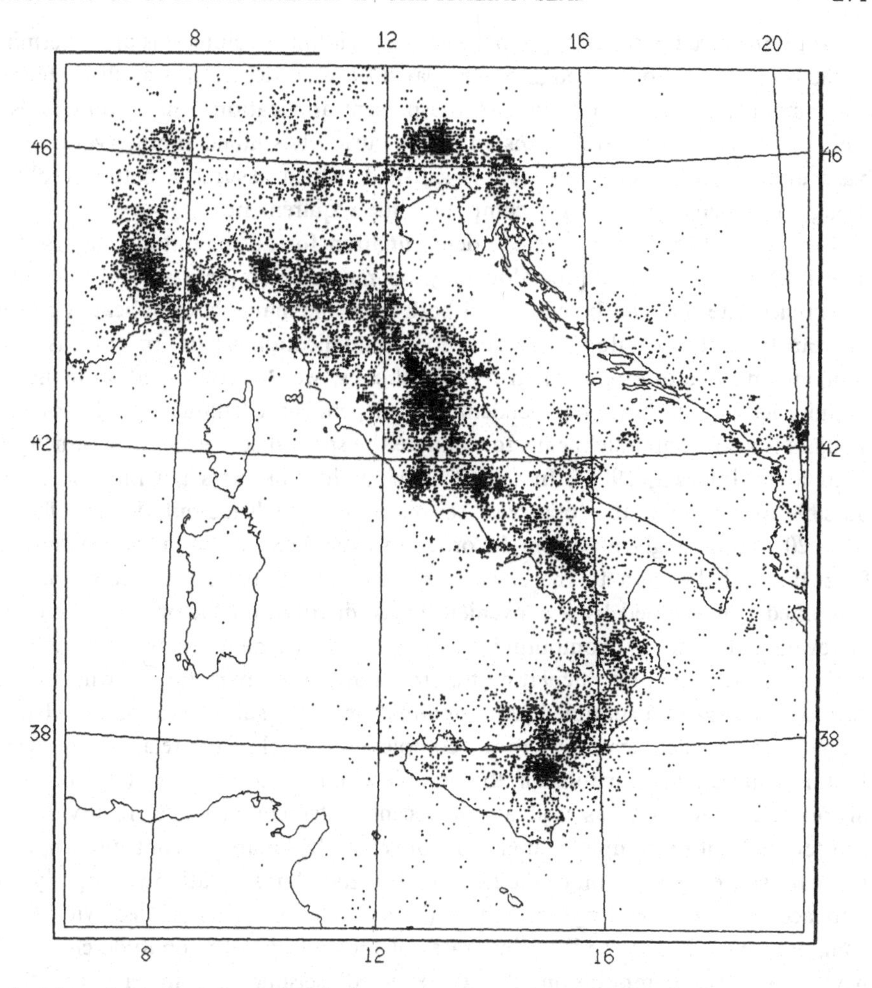

Fig. 2. Map of epicenters from PFG Italian earthquake catalog (1450 BC–1982). Only shallower earthquakes will be used in the analysis.

For the full description of the procedure, the reader should refer to the original paper quoted above, where it was presented for the first time and usefully applied to study seismicity in Calabria and Sicily. The method consists in a first phase of main-shock detection and aftershock removal, followed by a second phase in which each magnitude class sequence is repeatedly tested against poissonianity through a Kolmogorov–Smirnov one sample (KS1) test. The analysis leads to the partition of the total period spanned by the catalog into subintervals inside which the earthquake occurrence rate, apparent from the catalog, is constant and significantly different from the rate of the contiguous subintervals. For each magnitude class, the obtained apparent seismic rate is therefore a piecewise function of time from which the different completeness levels of the subintervals appear quite clearly. The variant introduced here refers to the way in which time subintervals are determined. In the original method, a rough binary partition was successively applied chiefly to accommodate algorithm rapidity. In the present application, time subintervals are built by trying a backward extension of the currently studied subinterval on a 5 yr basis until the KS1 statistical test is satisfied, which assures a better resolution, though at expenses of a larger computer time requirement.

One result of the analysis is the determination for all magnitude classes of the time interval in which the catalog may be reputed complete and of the related occurrence rate that can be computed both from the main-shock subcatalog as well as from the entire catalog, according to the application needs. In our case, since tsunamis may be even generated by aftershocks, we are concerned with the total seismic activity of both independent and dependent earthquakes. Completeness analysis is also a valuable means to evaluate the space distribution of seismicity (see Tinti and Mulargia, 1985). In fact, following the time axis partition stage, it is possible to determine the 'observed' seismic rate of each magnitude class for each $20' \times 20'$ cell of a given region. In order to extend the information on seismicity from the global region to the elementary cells, the global seismic rate previously computed is multiplied by a convenient space distribution factor. This factor may be evaluated first by computing a normalized seismicity map for each time subinterval and then by calculating the final normalized map for the whole catalog period by means of a suitable linear combination of the subinterval maps, where the weights are assigned according to the completeness level of the related subintervals. In this manner, all portions of the seismic catalog, i.e. the strongly incomplete historical first sections as well as the recent instrumental parts, relatively much shorter and more complete, make a conveniently balanced contribution to the resultant space distribution of the earthquakes and all information is properly taken into account. At this stage of the analysis, each cell is associated with the set composed by the 'observed' seismic rates of the considered magnitude classes. The next step is the computation of the 'expected' seismic rate in each cell for an arbitrarily selected magnitude interval based on a convenient frequency-magnitude relation. In this investigation, the Gutenberg–Richter exponential magnitude distribution is assumed for each cell, i.e.

$$N = N_a \exp[-\beta(M - M_0)],$$

where N is the annual expected number of earthquakes with magnitude exceeding M and M_0 is the lower magnitude threshold ($M_0 = 2.4$ in our study), while N_a and β are, respectively, the activity parameter and the decay parameter. Application of standard statistical methods enables us to infer both parameters together with their confidence bands and, in a following stage, to evaluate the mean number N of earthquakes per year with magnitude exceeding M for any cell. The calculation of both parameters N_a and β has been performed for all cells in the eight regions and represents the final step of the seismicity analysis.

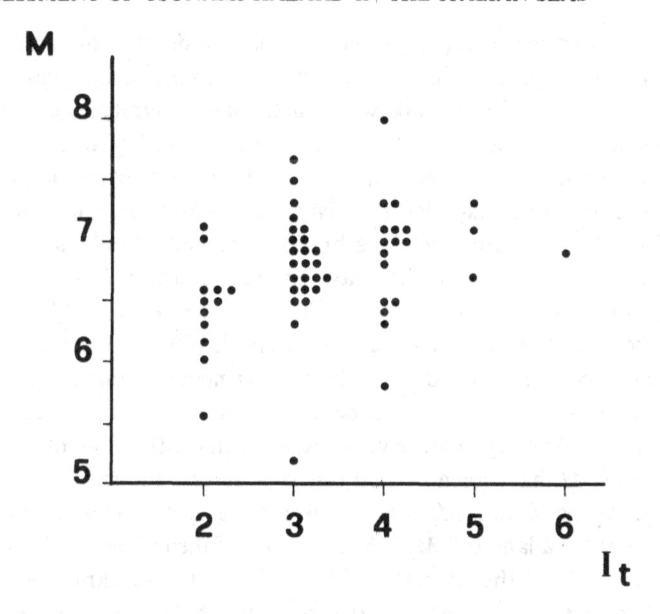

Fig. 3. Tsunamigenic earthquake magnitude is plotted against tsunami intensity, expressed in the European Soloviev–Ambraseys scale (see Ambraseys, 1962), for the tsunamis occurred in the seas surrounding Greece. Data taken from the Greek tsunami catalog by Papadopoulos and Chalkis (1984).

Precise theoretical relations between the earthquake focal characteristics and the ensuing tsunami are difficult to establish, since the generation mechanism is very complex, including a dynamic coupling between the earth system and the overlying water system, which cannot always be described by simple linear approximations or analytical means, especially in the very first nucleation phase. Attempts to correlate earthquake magnitude to tsunami size, either intensity or magnitude, led to expressions derived experimentally for the Pacific Ocean where tsunami catalogs contain more observations and data: for example, a class of these relationships concerns the minimum magnitude for a submarine earthquake to set in motion a tsunami (Murty, 1977; Iida, 1986). In the Mediterranean basin, the magnitude threshold M_t seems to be slightly smaller than in the Pacific, due perhaps to the shallower focal depth of the crustal generating quakes (see Soloviev, 1990; Tinti, 1990). Figure 3 shows, for example, a plot of earthquake magnitude vs. tsunami intensity for tsunamis which occurred around Greece, according to data from the Papadopoulos and Chalkis catalog (1984). It may be seen that tsunamis are usually produced by earthquakes with magnitude exceeding 6, though there are few examples of tsunamigenic shocks of lower magnitude. One of the important features of the tsunami sources in the Mediterranean is that most are very close to the coastline, frequently having epicenters some kilometers inshore or even further from the coast on land (see Figure 1 for the Italian tsunamigenic shocks). There is, therefore, a need to take into account the effect of the epicenter distance from the coastline of the nonsubmarine earthquakes on the magnitude threshold M_t to produce a tsunami. Though this is difficult to ascertain precisely, some clues may derive from the experimental relationships between earthquake magnitude and fault dimensions (see Aki and Richards, 1980). The idea is that an earthquake, with its focus beneath a land area and far from the sea, must involve a fracture surface large enough to cause the sea bottom to experience a displacement and to set long water waves in motion. Since the fault area increases broadly with magnitude, so larger earthquakes may excite tsunamis, even if they have epicenters further from the sea. Hence, M_t is reasonably an increasing function of the distance from the shore.

In the present application, the basic space elements are the $20' \times 20'$ cells, inside each of which seismic sources are assumed to exhibit homogeneous features: a unique M_t value is associated to each cell, with M_t depending on the distance of cell mid point from the nearest coastline segment. Several experiments with different values of the distance-dependent M_t have been carried out. The results shown in the next section refer to the following values: M_t is 6 for all cells covering either totally or marginally a marine area; for a land cell M_t is 6.5, 7, or 7.5, if the cell center distance from the sea falls, respectively, in the intervals 0–30, 30–50, and 50–75 km; whereas, for even larger distances, M_t is assigned the extreme value of 8, never attained by the known Italian earthquakes. It should be noted that the tsunamigenic magnitude threshold M_t may vary from one cell to another, which represents a powerful feature of the present method of tsunami potential evaluation. In this study, the position of the seismic source with respect to the sea has been used to determine M_t; but in future applications, M_t may be suited more closely to the specific features of the seismicity in each cell that are relevant to tsunami generation by including, for example, an M_t dependence on the mean focal depth or on the predominant focal mechanism of the earthquakes. Given M_t, the tsunami potential for a cell is easily evaluated by computing the number of the expected earthquakes with magnitude larger or equal to M_t on the basis of the Gutenberg–Richter frequency-magnitude law with parameters appropriately inferred for any specific cell. This has been performed for all cells comprised in the eight regions shown in Figure 1, which gives a complete picture of the tsunamis of any size expected to occur in a given time period in the seas surrounding Italy. It should be noted that the resultant tsunami occurrence rate is the average rate of the tsunamis generated locally inside each cell, since no account is taken here for tsunami propagation. Of course, the tsunamis observed in a given place are those produced locally as well as those generated by neighboring or remote sources. Future investigations on the hazard due to tsunami attacks against specific locations should combine the results of the present analysis on local tsunami potential with an analysis of remotely generated tsunamis.

3. Results and Discussion

Computation of the mean tsunami occurrence rate has been carried out in all regions shown in Figure 1 on a $20' \times 20'$ cell grid. Assumptions about the

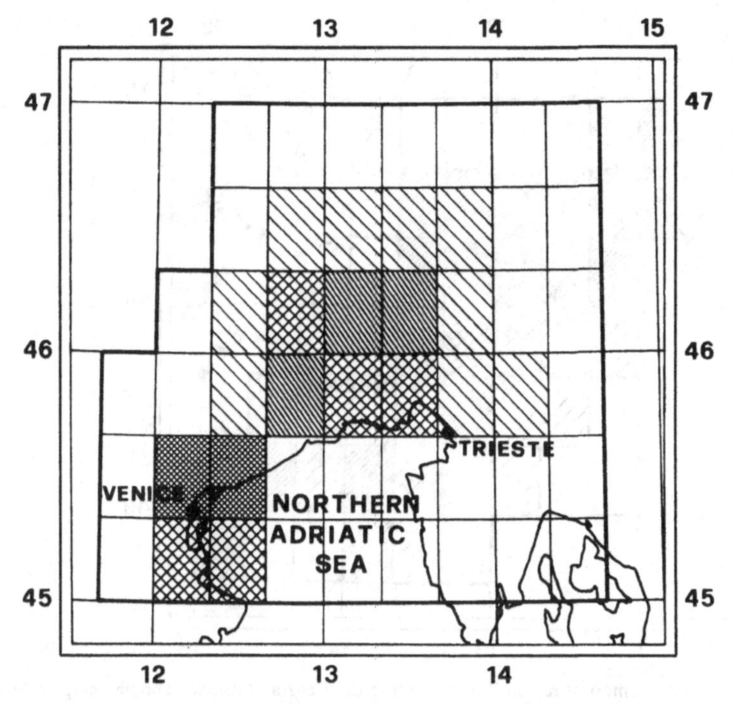

Fig. 4. Tsunami potential map of region 1, that is the Northern Adriatic. The mean number n of the expected, locally produced, tsunamis per 1000 yrs is given in a 5 deg scale, based on hatching density. Here the hatching code is the following one: white (w) $\rightarrow n \leqslant 0.01$, spaced oblique lines (sl) $\rightarrow 0.01 < n \leqslant 0.05$, slightly spaced oblique lines (ssl) $\rightarrow 0.05 < n \leqslant 0.10$, large squares (ls) $\rightarrow 0.10 < n \leqslant 0.35$, small squares (ss) $\rightarrow n > 0.35$.

tsunamigenic magnitude threshold M_t chosen in each cell were discussed in the previous section. The set of Figures 4–11 displays the tsunami potential maps. Expected tsunami rates are depicted through a graphic scale where higher density hatching corresponds to higher rate values, while white cells represent the smallest values and may be practically considered nontsunamigenic.

Figure 4 embraces the Northern Adriatic. It shows a rather low potential for tsunami generation, due to the substantial distance from the sea of the major seismic sources, which are responsible for some of the largest Italian earthquakes. The highest evaluated rates in the region are found on the western coasts close to the Venice Lagoon, where a diffuse seismicity may produce a mean number of tsunamis, say hereafter n, slightly larger than 0.35 per 1000 years. The resultant conclusion that tsunami hazard in Northern Adriatic is low agrees with the recent reexamination of the major events reported in the Caputo and Faita (1982, 1984) Italian catalog for this area. The revisions, in fact, corrected previous misinterpretations of the historical sources and led to a substantial reduction of the estimated tsunami sizes (see Guidoboni and Tinti, 1989).

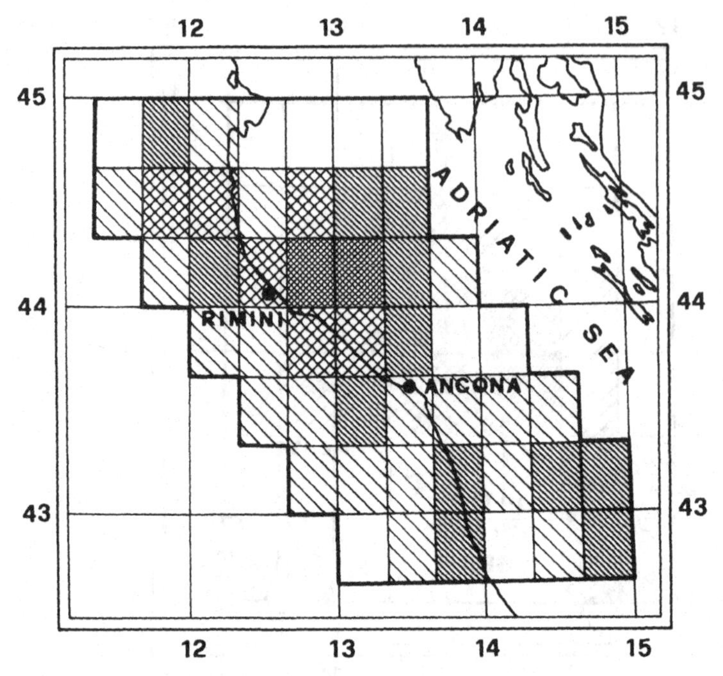

Fig. 5. Tsunami potential map of region 2, representing the Central Adriatic. The hatching scale used here for the expected tsunami rate n in a 1000 yr time basis is w $\rightarrow n \leqslant 0.01$, sl $\rightarrow 0.01 < n \leqslant 0.10$, ssl $\rightarrow 0.10 < n \leqslant 0.35$, ls $\rightarrow 0.35 < n \leqslant 0.70$, ss $\rightarrow n > 0.70$.

Figure 5 illustrates the Central Adriatic region, where seismic sources are active even offshore in a narrow belt parallelling the coastline. The highest rates exceed 0.7 tsunamis in 1000 years and are found in the cells around 44°N facing the southern Romagna flat beaches and the Northern Marche hilly coasts. Region 3, comprehending the Southern Adriatic as well as the Western Gulf of Taranto in the Ionian Sea, is displayed in Figure 6, where cells covering the Gargano promontory exhibit a potential higher than one tsunami per 1000 years, which is the largest found in the Adriatic Sea. One catastrophic tsunami is known to have occurred in this area in 1627, whose disastrous violence has been confirmed by recent critical review of all available coeval sources (Guidoboni and Tinti, 1988). These observations clarify that the Gargano generation region is one of the most interesting of the Italian peninsula and is certainly worthy of more detailed and intensified research.

Calabria and Eastern Sicily activity is shown in Figure 7, concerning region 4, which is the Italian area with the observed highest tsunami rate, being site of very active tectonic processes that imply both seismic energy release and active volcanism (see Tinti, 1989). Here some of the largest Italian tsunamis were generated: for example, the 1169 tsunami, which mainly affected Catania and the surrounding Sicilian coast, the events related to the Calabrian seismic crisis of the year 1783 (see Tinti and Guidoboni, 1988) and the 28 December 1908 tsunami. This tsunami, that

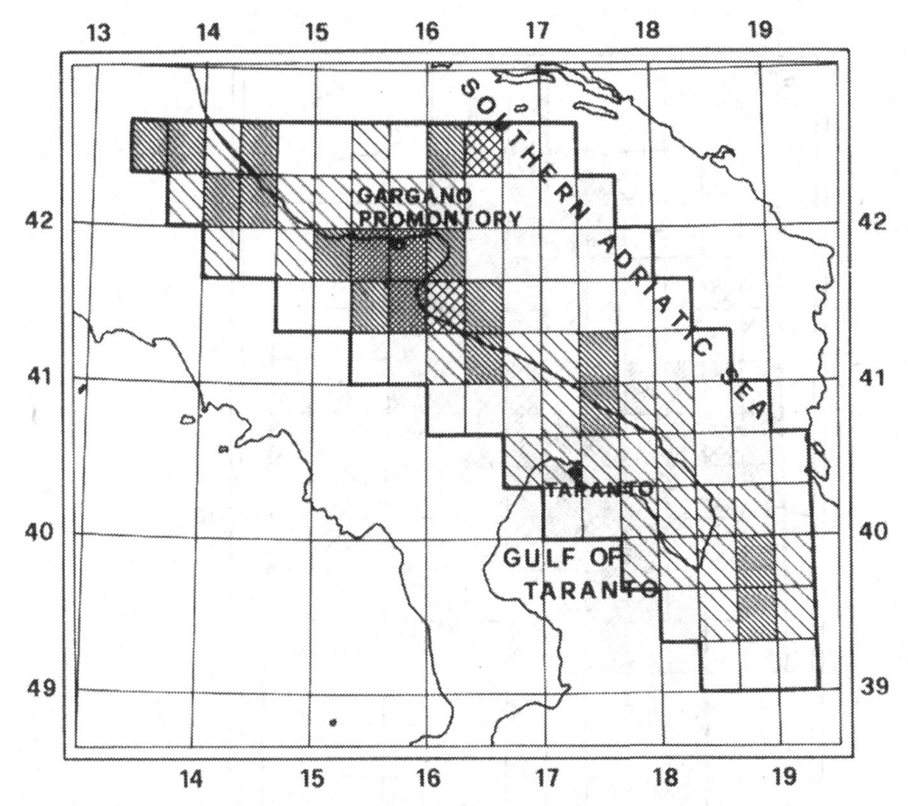

Fig. 6. Tsunami potential map of region 3: Southern Adriatic. The graphic code to represent n, i.e., the evaluated mean number of events in a 1000 yr period is w → $n \leqslant 0.05$, sl → $0.05 < n \leqslant 0.20$, ssl → $0.20 < n \leqslant 0.50$, ls → $0.50 < n \leqslant 0.90$, ss → $n > 0.90$.

invested the Messina Straits coasts and the Eastern Sicilian shores with a series of disastrous waves, was caused by a normal-fault $M = 7.1$ earthquake that lowered the bottom of the Straits. An appreciably good documentation is available for this tsunami, that has been the object of intense studies, including numerical simulations (see Tinti and Giuliani, 1983a, b; Barbano and Mosetti, 1983). In agreement with the historical observations, the calculated tsunami rate given in Figure 7 shows that the most relevant generation sources are placed in the Messina Straits and in the adjacent southern coastlines in the central part of Eastern Sicily: the value of n exceeds 10 tsunamis per 1000 years in the cell covering the Straits, whereas in three surrounding cells it exceeds 4 events per 1000 years. Further noticeable tsunamigenic sources may be found in the Southern Tyrrhenian Sea in a wide area including the Aeolian Islands, the Gulfs of Patti and of Milazzo in Northeastern Sicily and the Gulfs of Gioia and of Sant'Eufemia in Western Calabria. Furthermore, taking into account even the smaller sources in Northern (Gulf of Policastro) and in Eastern (Gulf of Squillace) Calabria, the total generation surface with local

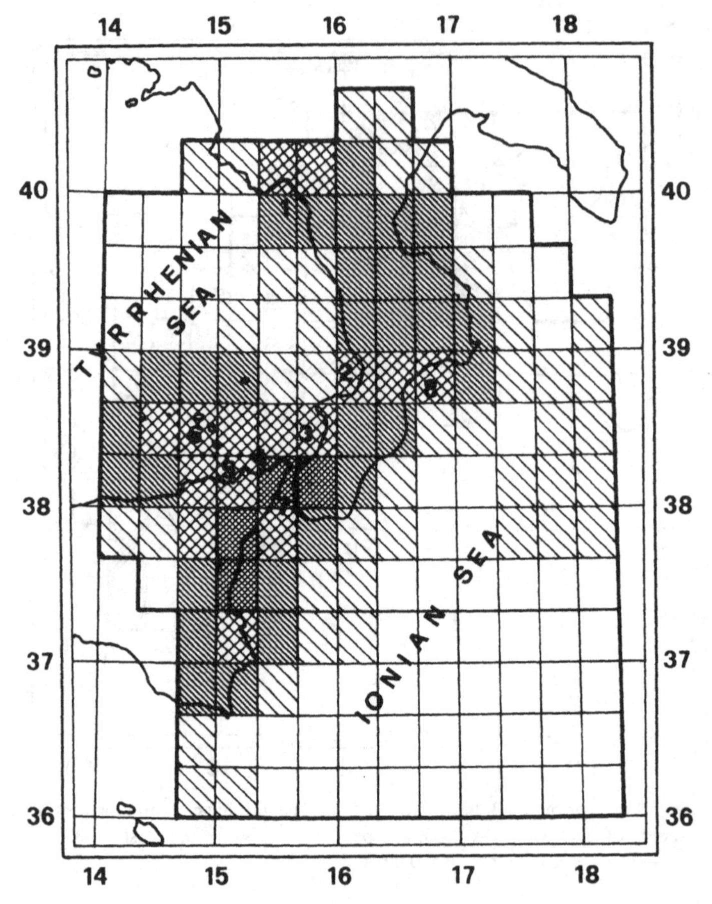

Fig. 7. Tsunami potential in region 4, displaying Calabrian Arc. It is the region with the highest calculated tsunami rate: the estimated potential of the cell including the Messina Straits exceeds 10 tsunamis in 1000 yr. The hatching meaning is w $\rightarrow n \leqslant 0.05$, sl $\rightarrow 0.05 < n \leqslant 0.50$, ssl $\rightarrow 0.50 < n \leqslant 1.5$, ls $\rightarrow 1.5 < n \leqslant 4.0$, ss $\rightarrow n > 4.0$. Numbers denote geographical features quoted in the text: 1 = Gulf of Policastro, 2 = Gulf of Sant'Eufemia, 3 = Gulf of Gioia, 4 = Gulf of Milazzo, 5 = Gulf of Patti, 6 = Aeolian Islands, 7 = Straits of Messina, 8 = Gulf of Squillace.

estimated potential larger than 1.5 events in 1000 years per cell includes 20 cells, which corresponds to an area of about 2×10^5 km^2.

Figure 8 refers to Western Sicily. Here the most relevant seismic faults are on land in the Belice Valley, where the major earthquake hitting the region was recorded in 1968 and did not cause any tsunami. The Italian catalog reports also tsunamis of volcanic origin in the Sicily Strait (see Figure 1) whose rate, however, cannot be estimated by means of the present method, based on seismicity analysis, as it was already stressed in the preceding section. The most important generation zone seems to be in the southern coasts, around the town of Sciacca, but the

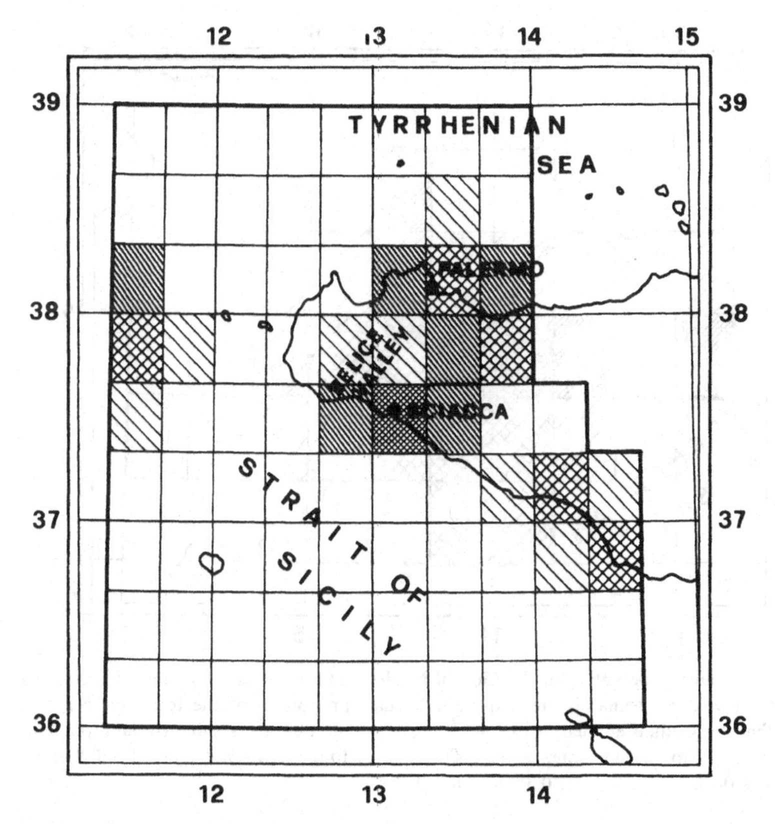

Fig. 8. Map of tsunami potential in region 5, Western Sicily. It is a region with a moderate tsunami rate, similar to regions 1 and 8. The hatching scale used in this figure refers to a 1000 yr period and may be described as w $\rightarrow n \leqslant 0.01$, sl $\rightarrow 0.01 < n \leqslant 0.05$, ssl $\rightarrow 0.05 < n \leqslant 0.10$, ls $\rightarrow 0.10 < n \leqslant 0.40$, ss $\rightarrow n > 0.40$.

expected tsunami rate is relatively small and comparable with the one calculated for the Northern Adriatic.

Figure 9 covers region 6, practically encompassing Campania and the facing Tyrrhenian Sea. The catalog reports here a relevant proportion of tsunamis due to the volcanic activity of Vesuvius. Tsunamis of seismic origin are relatively few and, therefore, it is not surprising the small tsunami rate resulting from our calculations. The main, though low-potential, tsunamigenic source embraces the Gulf of Naples and the opposite Islands of Procida and Ischia.

Region 7, encompassing the Latium and Southern Tuscany coasts, exhibits the smallest tsunami rate of all Italy, about one order of magnitude lower than in all other regions, as is shown in Figure 10. Reports of earthquakes go back several centuries BC in this area that was the core of the ancient Roman civilization, and completeness analysis reveals that the set of the large magnitude shocks is here complete for a period much longer than elsewhere in Italy. The long known history

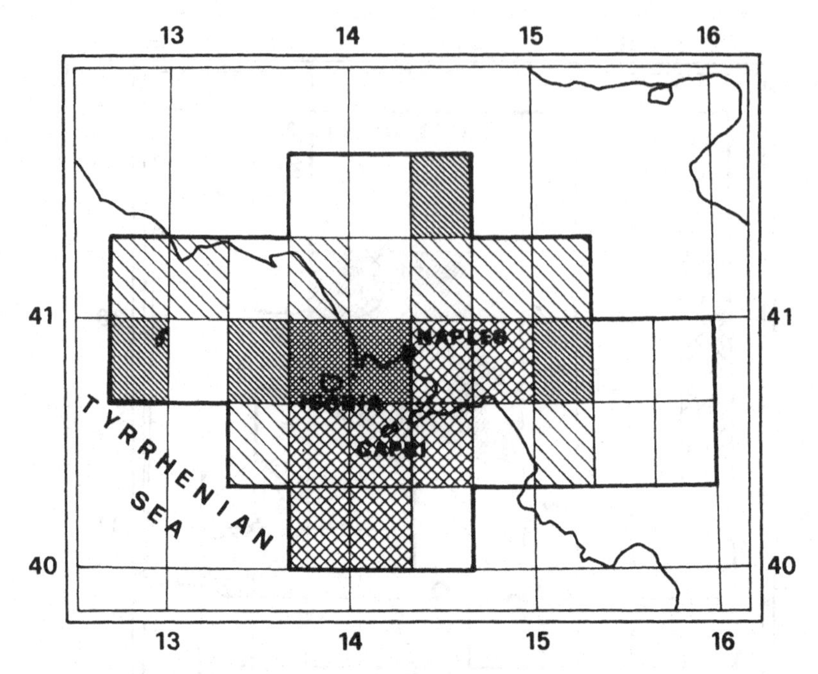

Fig. 9. Map of tsunami potential for the Gulf of Naples and the neighboring areas, forming region 6. Here reports of volcanic tsunamis are a quite important proportion of the total number of events. Remember that the method applied in this work may evaluate only the seismic tsunami potential. The graphic code, given for a time interval of 1000 yr, is as follows: w $\rightarrow n \leqslant 0.01$, sl $\rightarrow 0.01 < n \leqslant 0.05$, ssl $\rightarrow 0.05 < n \leqslant 0.10$, ls $\rightarrow 0.10 < n \leqslant 0.35$, ss $\rightarrow n > 0.35$.

of seismicity in the provinces around Rome discloses quite reliably that the main active seismic sources are both low potential and far from the sea, and therefore unlikely to give rise to substantial tsunami generation. The highest estimated tsunami rates moderately exceed one tsunami per 10 000 years; the total number of expected tsunamis generated in the region is much smaller than those that are reported in the catalog (see Figure 1). This discrepancy calls for a reexamination of the events reported to have occurred here. Future research should carefully investigate the reports' reliability in the attempt of reconciling estimates and observations. Apart from this misfit, however, this region must be considered an extremely low tsunami potential zone.

The last region of the partition studied, namely region 8, is comprehensive of the Liguria and Northern Tuscany coasts. The most energetic seismic sources are in Garfagnana at the northern end of the Tuscan Apennine, some tens of kilometers distant from the sea, whereas lower potential sources extend also in marine areas both to the east and to the west. The calculated tsunami rate n shows two distinct zones of higher expected activity, as is visible from Figure 11; but n is rather modest almost everywhere in the region, which may be consequently grouped with region 1 and region 5 in the areas of moderate tsunami potential.

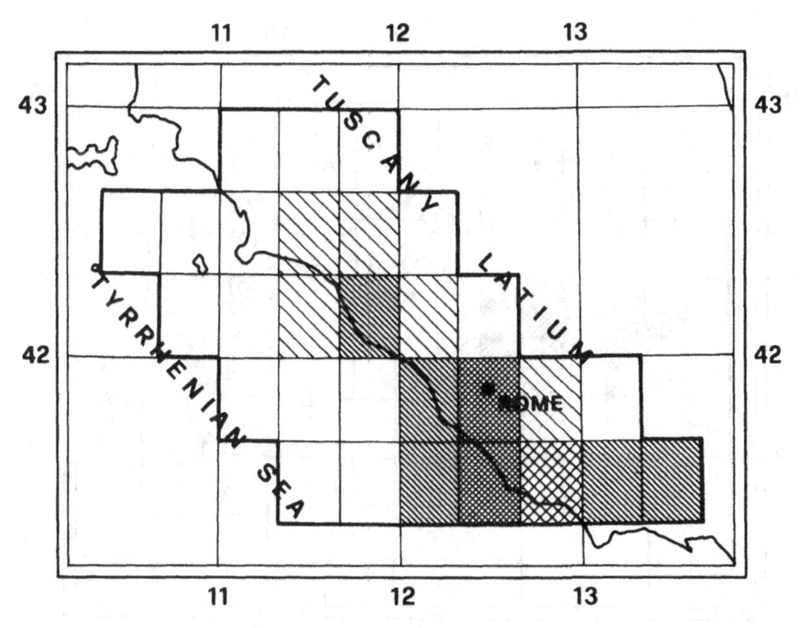

Fig. 10. Tsunami potential map in region 7, covering Latium and Southern Tuscany. Here the calculated rate values are extremely low and slightly smaller than those appearing from the Italian tsunami catalog. The hatching code, concerning a period of 10 000 yr, is the following one: w $\rightarrow n \leqslant 0.01$, sl $\rightarrow 0.01 < n \leqslant 0.10$, ssl $\rightarrow 0.10 < n \leqslant 0.35$, ls $\rightarrow 0.35 < n \leqslant 0.70$, ss $\rightarrow n > 0.70$.

The results obtained by means of the adopted method provide a rather detailed picture of the mean tsunami activity to be expected on the Italian coasts. The dimension of the basic space units taken in this investigation, i.e. of the $20' \times 20'$ cells, is a suitable compromise between the desired resolution in space and the requirements of the statistical analysis. Smaller elementary units would have pushed up to a critical point the evaluation of the space distribution factor, whose role in the procedure has been described in the previous section. But larger cells would have degraded the assumption of homogeneous seismicity inside each cell, which is needed to evaluate the minimum tsunamigenic magnitude M_t. One merit of this analysis is the rigorous and systematic application of a unique, clearly-defined method to all regions around Italy. This approach permits a fair comparison between regions and between cells belonging to the same region or to different regions, and a reliable ranking. It has been shown that the region with the highest tsunami potential in Italy is the Calabrian Arc including the Messina Straits and the seas around its northern and southern mouths in region 4. Second in the list comes the Gargano promontory zone in the Southern Adriatic, included in region 3. All disastrous tsunamis of seismic origin reported in the catalog pertain to these two regions. In the third place may be put the Central Adriatic province in region 2, followed by Northern Adriatic, Western Sicily, and Ligurian Gulf more or less exhibiting the same moderate level of expected tsunami rate. Last may be classified

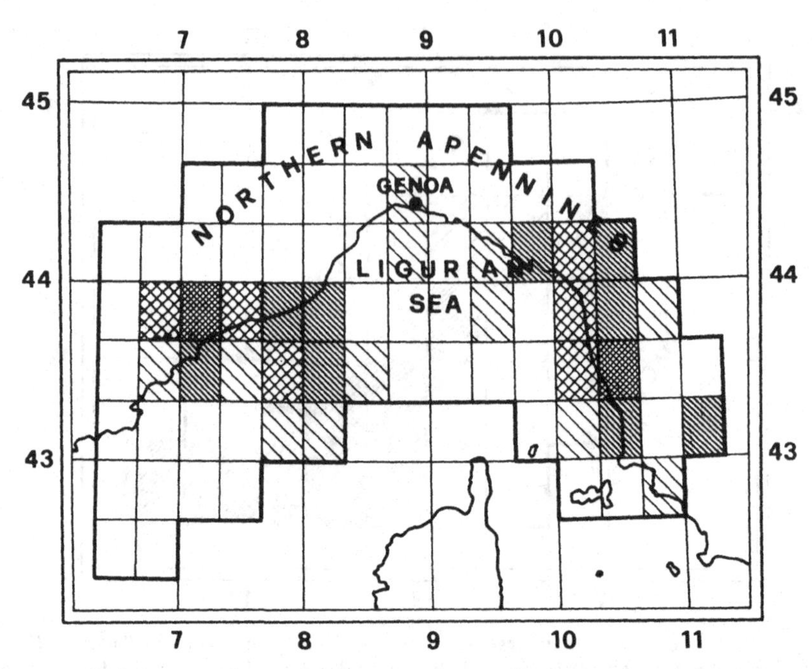

Fig. 11. Map of the tsunami potential for region 8, namely Liguria and Northern Tuscany. The tsunami rate expected on the basis of the calculations performed is rather low. The meaning of the various hatching patterns is given in the following and refers to a time period of 1000 yr: w $\rightarrow n \leqslant 0.03$, sl $\rightarrow 0.03 < n \leqslant 0.09$, ssl $\rightarrow 0.09 < n \leqslant 0.18$, ls $\rightarrow 0.18 < n \leqslant 0.35$, ss $\rightarrow n > 0.35$.

region 7, where tsunamis should be considered very rare events. Comparison with the reports included in the tsunami catalog shows a very good agreement between what was observed and what is computed on a completely independent basis, apart from the discrepancy already pointed out for region 7.

It is worth stressing once again the preliminary character of this investigation. All the results shown deserve to be refined. There are indeed some points of the analysis performed here that may be improved or modified. Some assumptions at the stage of the completeness analysis or of the subsequent statistical applications, such as the Gutenberg–Richter frequency-magnitude law, may be changed. Moreover, some more complicated criteria may be adopted in order to evaluate the capability of an earthquake to produce a tsunami, instead of the simple concept of the tsunami magnitude threshold M_t used here, as already discussed in the preceding section. Future studies should also incorporate nonseismic sources particularly active in region 6, and also in regions 4 and 5; the complexity of propagation and dissipation should be introduced as well in order to evaluate tsunami hazard on specific points on the coasts. All the above are only a few examples of future possible extensions and continuation of this work that must be considered as the first systematic approach to the tsunami potential evaluation in the Italian seas.

Acknowledgements

The work presented here was partially financed by the Italian CNR (Consiglio Nazionale delle Ricerche) and the MURST (Ministero della Università e della Ricerca Scientifica e Tecnologica). The author wishes to thank Mr Massimo Bacchetti for drawing most of the figures.

References

Aki, K. and Richards, P. G.: 1980, *Quantitative Seismology. Theory and Methods*, W. H. Freeman, San Francisco.

Ambraseys, N. N.: 1962, Data for the investigation of the seismic sea-waves in the Eastern Mediterranean, *Bull. Seism. Soc. Am.* **52**, 895–913.

Barbano, M. and Mosetti, R.: 1983, A hydrodynamical model of tsunami waves propagation in the Messina Strait, *Boll. Geofis. Teor. Appl.* **98**, 83–95.

Bedosti, B. and Caputo, M.: 1986, Primo aggiornamento del catalogo dei maremoti delle coste italiane (in Italian), *Atti Accademia Nazionale dei Lincei, Rendiconti, Classe Scienze Fisiche, Matematiche, Naturali, Serie VIII, Vol. LXXX*, 570–584.

Caputo, M. and Faita, G.: 1982, Statistical analysis of the tsunamis of the Italian coasts, *J. Geophys. Res.* **87**, 601–604.

Caputo, M. and Faita, G.: 1984, Primo catalogo dei maremoti delle coste italiane (in Italian), *Atti Accademia Nazionale dei Lincei, Memorie, Classe Scienze Fisiche, Matematiche, Naturali, Serie VIII, Vol. XVII*, 213–356.

Guidoboni, E. and Tinti, S.: 1988, A review of the historical 1627 tsunami in the Southern Adriatic, *Science of Tsunami Hazards* **6**, 11–16.

Guidoboni, E. and Tinti, S.: 1989, The largest historical tsunamis in the Northern Adriatic Sea: A critical review, *Science of Tsunami Hazards* **7**, 45–54.

Iida, K.: 1986, Activity of tsunamigenic earthquakes around the Pacific, *Science of Tsunami Hazards* **4**, 183–191.

Murty, T. S.: 1977, *Seismic Sea Waves – Tsunamis*, Bulletin of the Fisheries Research Board of Canada, Ottawa.

Papadopoulos, G. A. and Chalkis, B. J.: 1984, Tsunamis observed in Greece and the surrounding area from antiquity up to the present times, *Marine Geol.* **56**, 309–317.

Soloviev, S. L.: 1990, Tsunamigenic zones in the Mediterranean Sea, *Natural Hazards* **3**, 183–202.

Tinti, S.: 1989, Tsunami activity in Italy and surrounding area, in E. Boschi, D. Giardini and A. Morelli (eds), *A Mission to the Planet Earth*, Galileo Galilei, Rome.

Tinti, S.: 1990, Tsunami research in Europe, *Terra Nova* **2**, 19–22.

Tinti, S. and Giuliani, D.: 1983a, The Messina Straits tsunami of the 28th of December 1908: a critical review of experimental data and observations, *Nuovo Cimento* **6C**, 424–442.

Tinti, S. and Giuliani, D.: 1983b, The Messina Straits tsunami of the 28th of December 1908: An analytical model, *Ann. Geophys.* **1**, 463–468.

Tinti, S. and Guidoboni, E.: 1988, Revision of the tsunamis occurred in 1783 in Calabria and Sicily (Italy), *Science of Tsunami Hazards* **6**, 17–22.

Tinti, S. and Mulargia, F.: 1985, Completeness analysis of a seismic catalog, *Ann. Geophys.* **3**, 407–414.

Tinti, S., Vittori, T. and Mulargia, F.: 1986, Regional Intensity-Magnitude relationships for the Italian territory, *Tectonophysics* **127**, 129–154.

Tinti, S., Vittori, T. and Mulargia, F.: 1987, On the macroseismic magnitudes of the largest Italian earthquakes, *Tectonophysics* **138**, 159–178.

Natural Hazards **4**: 285–292, 1991.

Assessment of Project THRUST: Past, Present, Future

EDDIE N. BERNARD
Pacific Marine Environmental Laboratory, National Oceanic and Atmospheric Administration, Bin C15700/Bldg. 3, 7600 Sand Point Way N. E., Seattle, Washington 98115-0070, U.S.A.

(Received: 15 March 1990; revised: 18 June 1990)

Abstract. Project THRUST (*T*sunami *H*azards *R*eduction *U*tilizing *S*ystems *T*echnology) was a demonstration of satellite technology, used with existing tsunami warning methods, to create a low cost, reliable, local tsunami warning system. The major objectives were successfully realized at the end of the demonstration phase in September 1987. In June 1988, the Chilean Government held a workshop to assess the value of THRUST to national interests. Two recommendations came forth from the workshop: (1) the technology was sufficiently reliable and cost-effective to begin the development of an operational prototype and (2) the prototype would be used as the Chilean Tsunami Warning System. As of August 1989, the equipment was in operational use. In September 1989, major improvements were made in the satellite operations that reduced the response time from 88 to 17 sec and enlarged the broadcast area by 50%. The implications of the recent improvements in satellite technology are discussed for application to reductions in disaster impacts.

Key words. Tsunami, warnings, satellite communications, rapid-onset natural hazards.

1. Past – Development of a Local Warning System

Throughout history, natural disasters have killed and disrupted people of every nation on the globe. Rapid-onset natural hazards such as earthquakes, landslides, tsunamis, hurricanes, tornadoes, floods, volcanic eruptions, and wildfires have claimed more than 2.8 million lives worldwide in the past 20 years, adversely affecting 820 million people. The world's vulnerability and the social and economic cost of these hazards will only increase in the future due to population growth and urban concentration; increased capital investment coupled with new technologies; the existence of vulnerable critical facilities and fragile lifelines; and increasing interdependence of local, national, and international communities (Housner, 1987).

Tsunamis are one such rapid-onset hazard. Over 51 000 coastal residents have been killed by 94 destructive tsunamis in the past century. Lander and Lockridge (1986) have found that 99% *of all tsunami-related fatalities occur within 400 km of the earthquake epicenter*. Since tsunamis are generated over the area of uplift (which could be up to 600 km), most of the fatalities probably occurred on the coasts directly opposite the source with travel times less than 10 min.

Existing tsunami warning systems are effective on a *Pacific-wide* time scale of 1 h and a *regional* time scale of 10 min (Sokolowski, 1990). They are not effective, however, on a *local* time scale – i.e., within 2 min of a local, potentially tsunamigenic earthquake (Bernard *et al.*, 1988). Project THRUST has successfully designed and developed a warning system to meet this need by establishing a pilot system in Valparaiso, Chile. The three major components that make up the THRUST system's approach to tsunamis hazard mitigation on the local level are

- Pretsunami preparedness, consisting of historical data base studies, numerical model simulations, and development of a detailed emergency operations plan
- Real-time local hazard assessment, achieved by utilizing seismic triggers with predetermined threshold levels
- Rapid dissemination of information to local officials, achieved by exploiting satellite communications technology

The satellite-based communication system is shown schematically in Figure 1. A threshold detector sensor (A) provides a signal in response to a change that exceeds a preset level. (Sensors can monitor seismic activity, water heights, nuclear radiation, chemical concentrations, or any other activity for which a threshold detector is available.) The sensor signal activates a satellite uplink transmitter (B) into a preprogrammed transmission sequence. The transmitted signal is a unique address that identifies the sensor and transmitter. The GOES (Geostationary Operational Environmental Satellite) satellite (C), which resides in geosynchronous orbit, relays the coded message to the ground station (D). The ground station computer (E) processes incoming signals and decodes the message. The computer is programmed to react to the unique signal by commanding the ground station to transmit a short coded message (related to the sensor) to the satellite. This coded message is broadcast through the satellite and is received at the alert station (F). The alert station is a low-cost receiver and antenna tuned to the downlink frequency of the satellite and a decoder and processor that continuously monitor incoming data. The processor continuously scans all incoming messages from the downlink receiver. The processor is programmed to respond upon receipt of specific messages. The

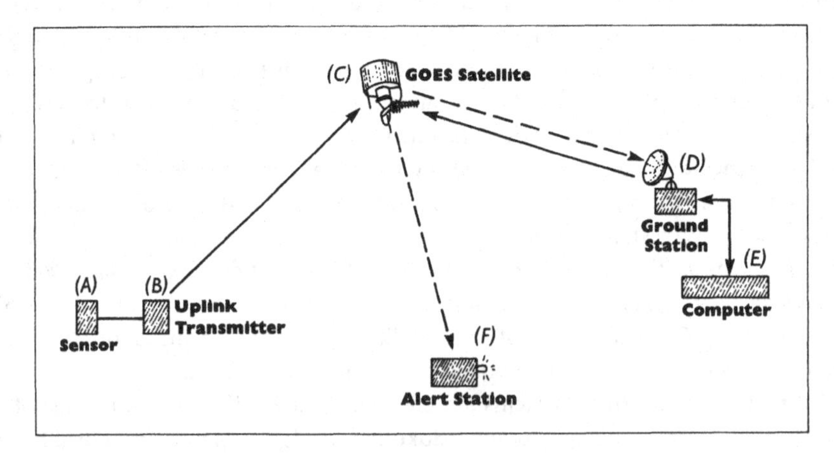

Fig. 1. Satellite communication system used for THRUST project.

response is typically a printed alert message indicating the time and location of the sensor trigger and action that should be taken by the owner of the receiver. The processor can also activate lights, acoustic alarms, telephone dialers, and other emergency responses.

After the pilot system was installed in September 1986, Project THRUST researchers began a lengthy evaluation program. Over a period of 9 months, the average response time of the communications technology was found to be 2 min. Transmission performance tests conducted during this period also demonstrated that the GOES communication system could be used in a warning mode. The THRUST team found that after one year of operation, hardware reliability was acceptable for first-generation pilot instrumentation.

The major objective of project THRUST had thus been achieved – the development of a *low-cost* system to deliver tsunami warnings useful on a *local* level. Hardware for the most basic THRUST system configuration, consisting of a seismic station and a tsunami warning station, cost about $15 000. Furthermore, the fundamental system delivered the primary THRUST product – life-saving tsunami hazard information – in an average elapsed time of 2 min simultaneously to anyone under the communication umbrella (Figure 2) with the appropriate receiver.

2. Present – Acceptance of THRUST Technology into Warning Operations

In June 1988, the Instituto Hidrografico de la Armada de Chile (IHA) hosted a THRUST workshop to evaluate the use of this early warning technology. The three-day workshop was attended by 50 scientists, emergency experts, and

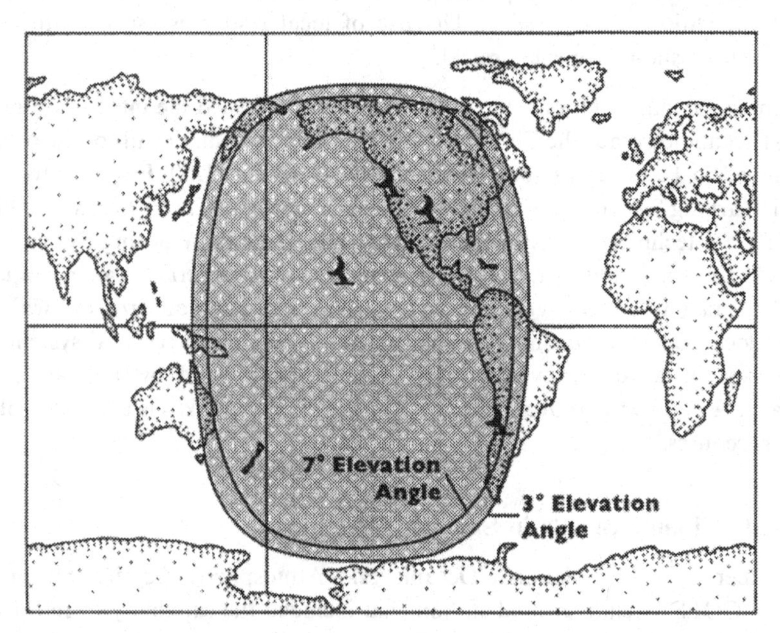

Fig. 2. Communication coverage of the GOES west satellite and locations of THRUST receivers.

emergency administrators from seven American countries. Two days were dedicated to the technical details of the hazard assessment (historical data analysis, inundation modeling, and emergency operating plan development), the equipment (engineering development, satellite operations, and performance evaluation), and the scientific methodology (seismic threshold setting, flooding maps) used in the project. The third day was used for demonstrations of the system and a session to formulate recommendations and conclusions (Espinosa, 1989).

The participants adopted the following recommendations at the end of the THRUST workshop:

- The THRUST pilot project demonstrated that the technology exists to develop an operational system. It is recommended that a two-step approach be adopted for the operational development:

  Step 1: Use existing equipment to instrument more sites of potential large earthquakes.

  Step 2: Write specifications for second-generation equipment based upon deficiencies found in the pilot project.

- Local emergency authorities should be encouraged to reexamine numerical simulations for a wider range of most probable earthquakes.
- Local emergency authorities should be encouraged to maintain close ties with the tsunami and seismological research communities to assimilate new technologies into warning, planning, and relief efforts.
- The capability for planning and carrying out numerical simulations should be transferred to local hazard planning and management authorities as efficiently and expeditiously as possible. The use of local resources (such as personal computers) should be encouraged.

With these recommendations, the experts at IHA continued the integration of the THRUST technology into the Chile Tsunami Warning System. Details of the process are explained in E. Lorca's paper, 'Integration of the THRUST Project into Chile Tsunami Warning System', in this special issue of *Natural Hazards*. Lorca concludes that THRUST technology has improved the Chile Tsunami Warning System.

It should be noted, however, that the system has yet to experience an earthquake of sufficient magnitude to trigger the system. This is a common problem with the testing of local/regional tsunami warning systems. The Hawaii regional system, for example, was installed for over 4 years before it was jolted by a magnitude 6 earthquake. A good way to overcome the absence of earthquakes is to devise robust testing procedures.

3. Present – Improvements in Satellite Operations

In September 1989, the National Oceanic and Atmospheric Administration upgraded the GOES satellite ground station that included the capability to retransmit

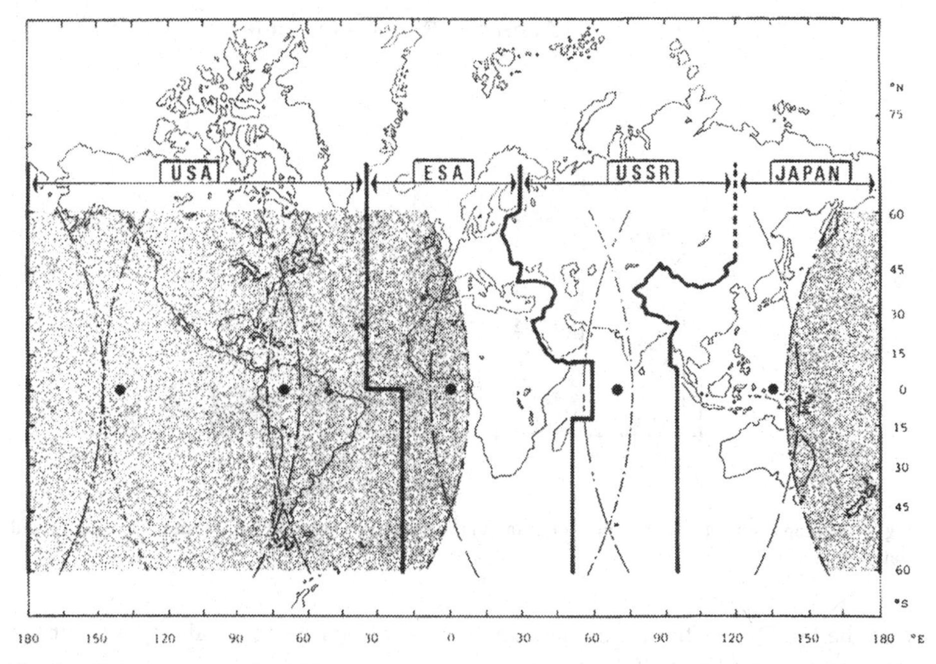

Fig. 3. Global coverage of GOES communication footprints. Darkened areas have THRUST emergency response capability.

messages such as those used in the THRUST project. The new satellite ground station, with its more advanced computer, significantly decreased the time of message processing. A special feature of these 'trigger' messages is a broadcast over both the NOAA GOES east and west satellites, which increases the area of broadcast by approximately 50% (Figure 3). This improved response time and broader coverage (about 62% of the surface of the Earth), along with low-cost, low-maintenance communication equipment makes the case for satellite-based, local early warning systems very appealing.

To quantify the improvements of the satellite ground station upgrades, two communication experiments were conducted – one before the upgrade and one after the new groundstation was operational. Each experiment was conducted as a series of communication tests. The first experiment was conducted from 1 March 1989 to 30 May 1989. The communication tests were designed to exercise the satellite system as if a real earthquake had occurred. The system was tested twice each week. One test was conducted during peak load (Monday 1300–1700 and Friday 1700–2200 GMT) while the other test was at random during that week. For the first experiment, a total of 44 tests were conducted with an average response time of 88 sec and a median response time of 47 sec. Eighty-four percent of the tests were executed properly. Of the seven failures, three occurred at peak load while five of the failures occurred at 1300 GMT (0800 local time). This experiment indicated

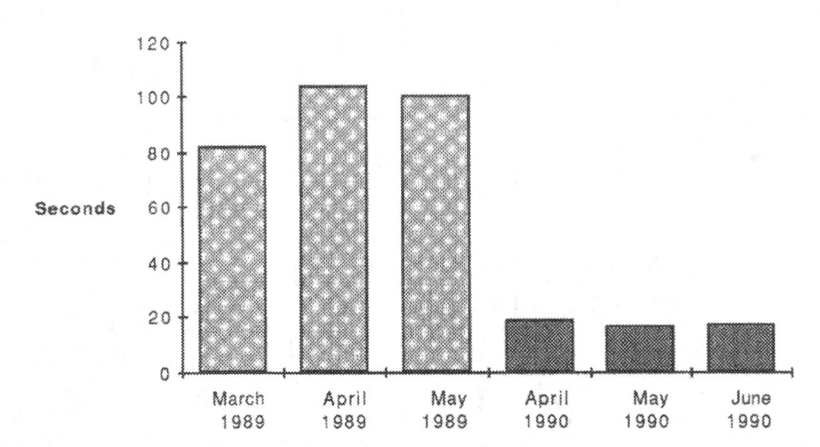

Fig. 4. Comparison of response time of communication tests conducted before and after the ground station upgrade.

that the GOES satellite communication system response and reliability were about the same in 1989 as in 1986.

The second experiment was conducted after the new satellite ground station was operational during April, May, and June 1990. The procedures in the first experiment were repeated as carefully as possible. During the second experiment, 23 tests were conducted with an average response time of 17 sec. All of these tests were executed properly. This experiment indicated that the changes in the GOES system have improved the response time by a factor of 5 (Figure 4).

The improved satellite system has an additional feature that tracks the time of receipt of a 'trigger' message and the time of transmission. For all the tests conducted during the second experiment, the satellite processing time was 1 sec. That is, the time the GOES satellite received the THRUST transmission to the time the GOES satellite system processed the message and retransmitted the response was 1 sec. The delays greater than 1 sec were due to processing within the THRUST equipment. With some engineering design work, it seems feasible that the response time could be reduced even more. At this stage of development, response times of 5–10 sec appear to be within technical reach. This signal tracking feature is also helpful in diagnosing problems in the system. A malfunctioning receiver was identified by tracing the signal through the system in November 1989. Upon replacement of the receiver, the system functioned properly.

4. Future – Additional Applications of THRUST Technology

After 4 years of testing and improvements, the THRUST-type system's approach to hazard mitigation appears to work quickly (17 sec) and reliably (greater than 90%).

Application to other hazards would be straightforward and should include preevent preparedness, real-time local hazard assessment through environmental monitoring, and rapid alert through satellite communications. Of these three, the satellite communications offer the broadest application. THRUST experience has demonstrated that satellite communications is the preferred way to alert populations of hazards. Satellite communications are fast, reliable, inexpensive, require low power (battery backup power is practical), and the satellite is removed from the local disaster (making it durable). These qualities are desirable for disaster mitigation.

Unlike conventional systems, this satellite-based alerting system does not rely on telephone line and microwave links to connect sensors to people who must respond to the sensor activity. Instead, low-powered and inexpensive sensor transmitters utilize a satellite and its operational support system to relay coded information in a near-real-time mode. Communications with the satellite can be maintained during periods of power outages, ground motion, and local disturbances that cause conventional communication systems to fail at critical times. The large broadcast area of the satellite coupled with the low cost of transmit and alert stations makes it possible to locate sensors in multiple hazard locations and place numerous alert stations in potentially affected areas for fast, effective response. The sensor and alert placement can be within a few feet of each other or as much as 12 000 miles (or any distance in between). Conventional alert systems require extensive human intervention and costly installations to provide similar broad geographical coverage. In addition, the nature of conventional communication systems is to transmit through a network of retransmission units that deliver information in a serial fashion. The satellite-based system broadcasts the same information from space such that all receivers are alerted at the same time.

5. Summary

The chronology of Project THRUST can be summarized as follows:

- 1982 – initial ideas
- 1984 – project begins
- 1986 – equipment installed and testing begins
- 1987 – project 'officially' ends (see Bernard *et al.*, 1988)
- 1988 – evaluation workshop (see Espinosa, 1989)
- 1989 – equipment in use as operational system (see Lorca, 1990)
- 1990 – improvements in satellite operations (average response time decreases from 88 to 17 sec)

The string of successes for the project in this chronology indicates the robustness of satellite communications. The fast response time makes this system attractive for alerting people for rapid onset disasters (such as earthquakes, landslides, flash floods, volcanoes, and even man-made events like nuclear power plant accidents and toxic material releases) that can be monitored by detection instruments. The broad geographical coverage makes the system desirable for alerting interested parties at near or far distances from the activity for taking evasive emergency action.

Acknowledgements

The author gratefully acknowledges the support of Mike Nestlebush of NOAA's National Environmental Satellite and Data Information Service and Emilio Lorca of IHA. Without their careful attention in testing the system, this research would not have been possible.

References

Bernard, E. N., Behn, R. R., Hebenstreit, G. T., González, F. I., Krumpe, P., Lander, J. F., Lorca, E., McManamon, P. M., Milburn, H. B.: 1988, On mitigating rapid onset natural disasters: Project THRUST, *Eos Trans., Amer. Geophys. Union* **69**(24), 649–661.

Espinosa, F.: 1989, Proceedings of THRUST Workshop, Instituto Hidrografico Armada de Chile.

Housner, G. W.: 1987, *Confronting Natural Disasters: An International Decade for Natural Hazard Reduction*, National Academy Press, Washington, DC.

Lander, J. F. and Lockridge, P. A.: 1986, Uses of a tsunami data base for research and operations (abstract), *EOS Trans., Amer. Geophys. Union* **67**, 1003.

Lorca, E.: 1990, Integration of the THRUST Project into Chile Tsunami Warning Systems, *Natural Hazards* **4**, 293–300 (this issue).

Sokolowski, T. J.: 1990, The Alaska Tsunami Warning Center, NOAA TM NWS AR-38, National Weather Service, Anchorage, AK.

Natural Hazards **4**: 293–300, 1991.

Integration of the THRUST Project into the Chile Tsunami Warning System

E. LORCA
Instituto Hidrografico de la Armada, Casilla 324, Valparaiso, Chile

(Received: 30 October 1989; revised: 25 June 1990)

Abstract. Several seismic gaps are identified along the coast of Chile. The main one is located in the northern part of the country (18° S–20° S), where a THRUST seismic trigger has been installed; two more triggers will be placed in other seismic gaps. The Standard Operations Plan (SOP) was tested during a disaster simulation exercise based on a major earthquake and tsunami situation. Response actions by governmental and civil agencies were monitored and a performance evaluation was done. Modifications of the SOP were found necessary to adjust the interactions between agencies.

Key words. Thrust, warning, Chile, tsunami.

1. Introduction

The operation of the National Tsunami Warning System (NTWS) has greatly improved over the past 5 years, although, however, the obtained level of activation time could not be reduced to less than 30 min after the tsunami generation. The THRUST system now operating in Chile, with seismic triggers in two different coastal places, can deliver seismic information from those specific areas within seconds of the earthquake's occurrence, allowing the NTWS to disseminate early warnings. More seismic sensors will be installed to cover the main known seismic gaps.

2. Tsunami Threat in Chile

Seismic potential for major and great earthquakes in Chile is directly related to the subduction of the Nazca plate below the South American plate. There are substantial variations in the seismicity along the Chile trench, resulting in seismic zones with different convergence rates. Several authors (Kelleher, 1972; McCann *et al.*, 1978; Nishenko, 1985; Stauder, 1973) have attempted the seismic zoning of the Chilean margin, recognizing areas with different seismic potential.

Results have been only rarely similar but all of them have one conclusion in common, i.e. the northern Chile area has a poorly known but significant amount of seismic potential. This region, sometimes called the 'Big Bend' area (18° S–20° S), has a history of destructive earthquakes and tsunamis (Lockridge, 1985), the last one in 1877, and is considered a candidate for a major or great earthquake in the

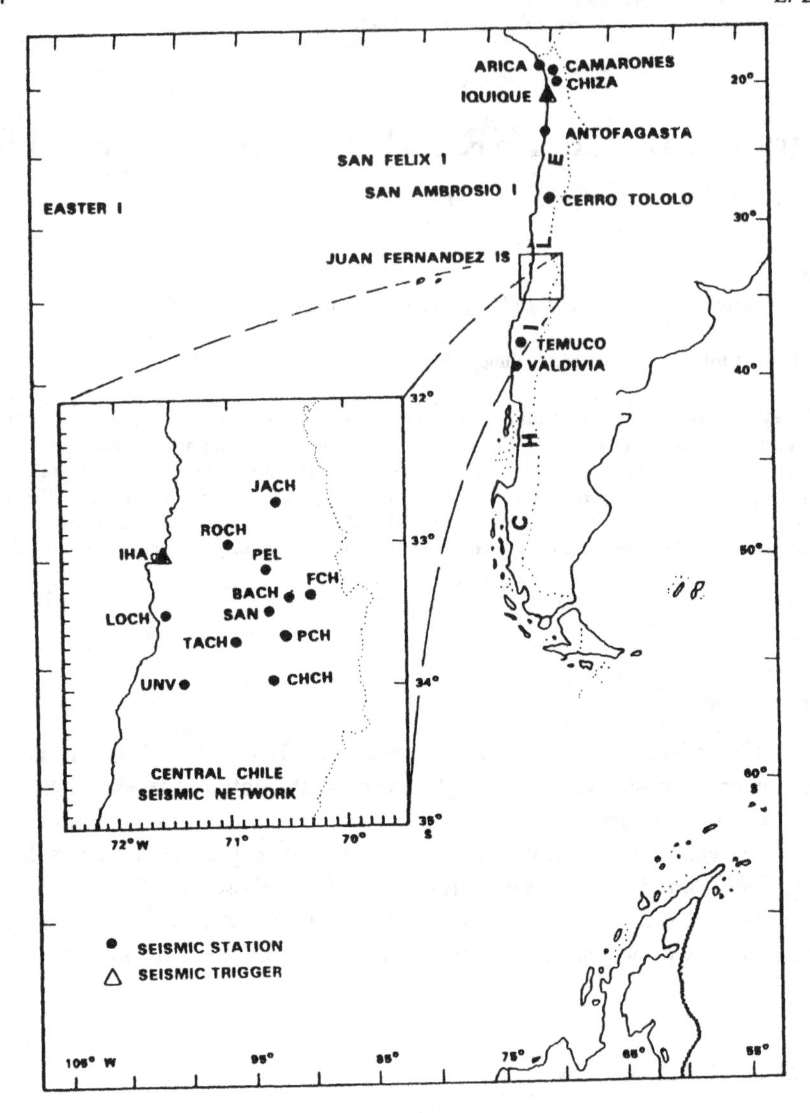

Fig. 1. Seismic network.

near future. A tsunami risk estimation (Lorca, 1985) shows that a maximum impact for local tsunami will affect this area, with expected runups of 10 m or more.

3. Seismic Coverage in Chile

Recent efforts have been made to improve the existing seismic network in Chile; Figure 1 is the present configuration which does not show a satisfactory coverage, except for the central part of the country. Late efforts were made to cover the northern part with four short-period seismographs with telemetering capability to Santiago.

The lack of a 24 hours-a-day operative capability of the university staff who run the seismic network is a serious drawback for the National Tsunami Warning System; in this case, the PTWC seismic location accuracy for Chile is improved by the Valparaíso seismic station data which can be retrieved at any time of the day, since the station is run by the Hydrographic Institute (IHA). However, PTWC magnitude and hypocenter determination is not received in Chile until 30 min after the earthquake origin time. This is the time frame that is covered with the THRUST seismic triggers, whose early warning capabilities were tested during the last year of the pilot project. One of these triggers has been installed in Iquique, 800 nautical miles north of Valparaíso, with an acceleration threshold of 0.126 g, while the second is located at IHA in Valparaíso with the same acceleration threshold. Those triggers will activate the GOES transmitters whenever there are earthquakes bigger than 6.2 Ms at 33 km from the sensor or 7.0 Ms at 80 km from the sensor (Bernard *et al.*, 1988).

A four-month testing of the operation of the system in 1989 shows an average response time of 88 sec with a minimum of 18 sec and a maximum of 6 min 15 sec.

Future plans for a THRUST system are to install a third trigger in another seismic gap and to move the one in Valparaíso to a southern location.

The information received from the seismic triggers is integrated with the normal seismic information received by the Hydrographic Institute, like that obtained from the IHA seismograph and the intensity data collected by the National Emergency Office (ONEMI). Within 15 min, travel times are calculated, a Tsunami Watch Bulletin is released to all the emergency agencies, as shown on Figure 2, and IHA seismic data is sent to the Pacific Tsunami Warning Center (PTWC).

4. Sea-Level Data

During the pilot project, the accelerometer-to-water level sensor system proved to be the weakest link of the three THRUST subsystems. The performance for this subsystem was only 78.9%, due to extended periods of time of internal power failures. For this reason, this subsystem has not been integrated into the NTWS and the instruments are being used as a backup for the existing system. However, the existing Chilean tide network (Figure 3) has 5 HANDAR DCP's with tsunami self-activation capacity whose data can be rapidly retrieved by PTWC. Notwithstanding the former, they are as far as possible away from the possible location of an impending earthquake in the northern part of the country; the existence of the THRUST water-level sensor in Iquique or any other seismic gap would make a perfect complement to the tide gauge network.

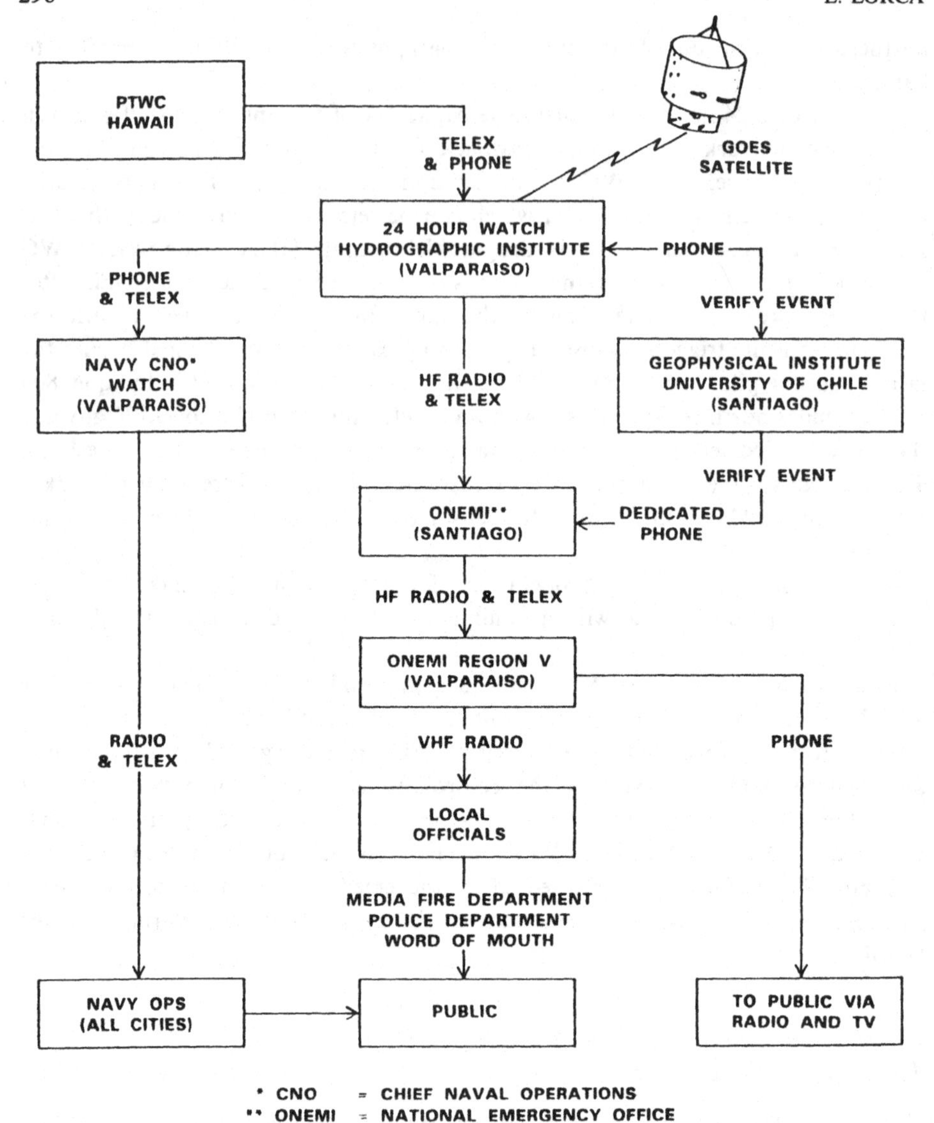

Fig. 2. National Tsunami Warning System Communications.

5. Standard Operations Plan

Modeling techniques were used by the THRUST team to estimate the inundation areas in Valparaíso. The selected model was previously used to evaluate tsunami threats in other areas of South America (Hebenstreit and Whitaker, 1983). The results of the numerical simulations, namely simulated inundation maps, were combined with baseline topography and street maps to identify security areas and

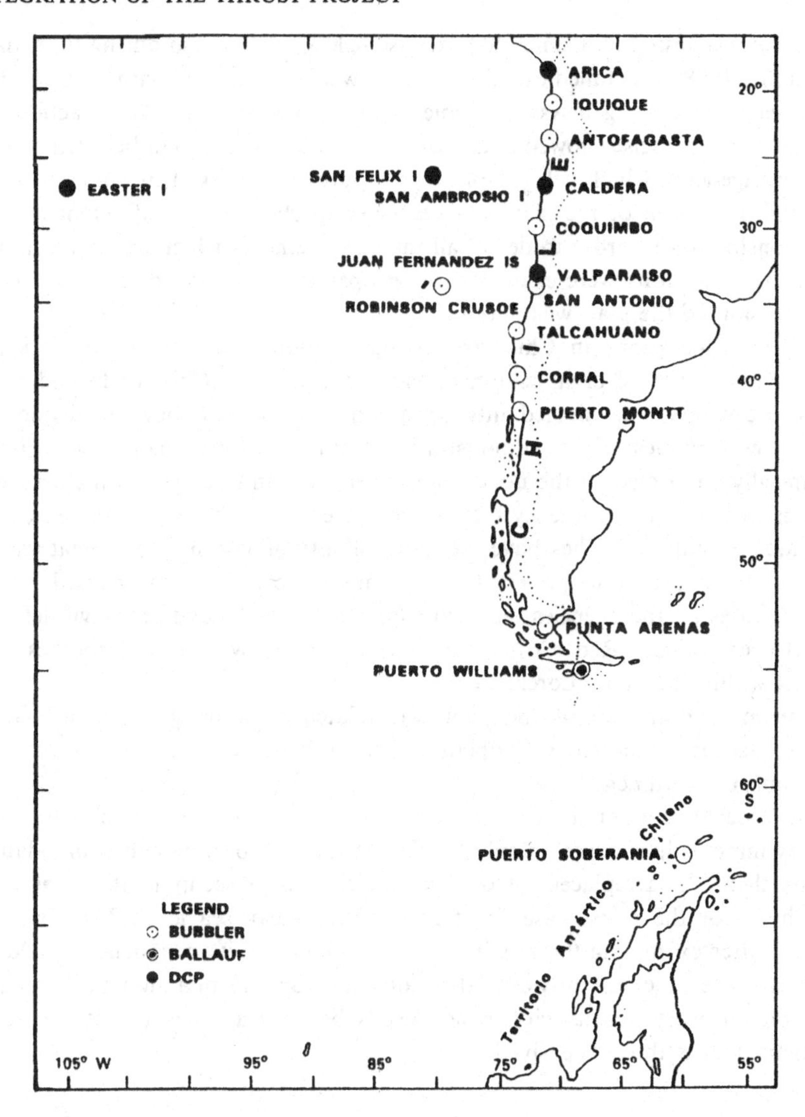

Fig. 3. Tide gauge network.

evacuation routes. This information was the basis for developing the 'THRUST Project Tsunami Emergency Operations Plan for Chile' (Lorca, 1987), which includes maps depicting probable inundation areas, evacuation routes, security areas, hospitals, short-term measures to be taken on issuance of a tsunami warning and longer-term relief efforts once the tsunami has receded, including responsibilities and functions of every disaster agency involved in a tsunami emergency.

This plan was tested in a simulated exercise held in Valparaíso during three days in December 1988. A detailed exercise scenario was implemented with input from a control team, generating news or problems to the participants. The reaction for them during the exercise showed a lack of coordination of the plan between several emergency agencies, which delayed proper measures being taken by some of them.

A detailed revision of the plan was necessary to check on the different actions and responsibilities in order to detect all the disagreements which arose during the exercise. Modifications were examined for proper inter-agency adjustments and a revised version of the plan was adopted.

Emergency operations in Chile are organized along the lines of the existing administrative structure organization of the country. Thus, Chile is divided in 13 regions, every region is divided into several provinces, and they are divided in communities (counties). Every administrative level has an Emergency Office (Figure 4) technically connected to the upper and lower level and dealing with emergency situations through an Emergency Operating Center (EOC) where all the executive phases are concentrated. They have the responsibility of assessing the threat and of taking action according to pre-established plans for every existing hazard. Every EOC integrates all the agencies that can support disaster operations: civil defense, armed forces, police, Red Cross, etc., and everyone with responsibilities and functions within the plan (Lorca, 1987).

The main modification of the plan was related to moving the coordinating response, assistance, and recovery operations from the regional level to the community (county) government level.

Several coastal communities in Chile will make use of the present plan with necessary minor adjustments, mainly relating to the lack of a modeling inundation study on them. As a replacement of this deficiency, a preset inundation has been fixed which, for the worse case, is at 25 m above mean sea level. This figure is arrived at after examining the existing historical data and its confidence should be revised. However, the best answer to the problem is to perform a numerical tsunami simulation on every coastal city, which needs better than the currently existing bathymetry and is thus expensive.

6. Conclusions

- The THRUST project products have been effectively integrated into the operations of the National Tsunami Warning System, thus allowing a faster and better response from the disaster agencies.
- THRUST seismic triggers are very good tools to complement any existing network, since they can give preliminary location and magnitude estimations very rapidly, so every seismic gap should be covered by them.
- The second generation THRUST instrumentation, including the water-level capabilities, should be implemented as soon as possible in order to make a wider use of them.

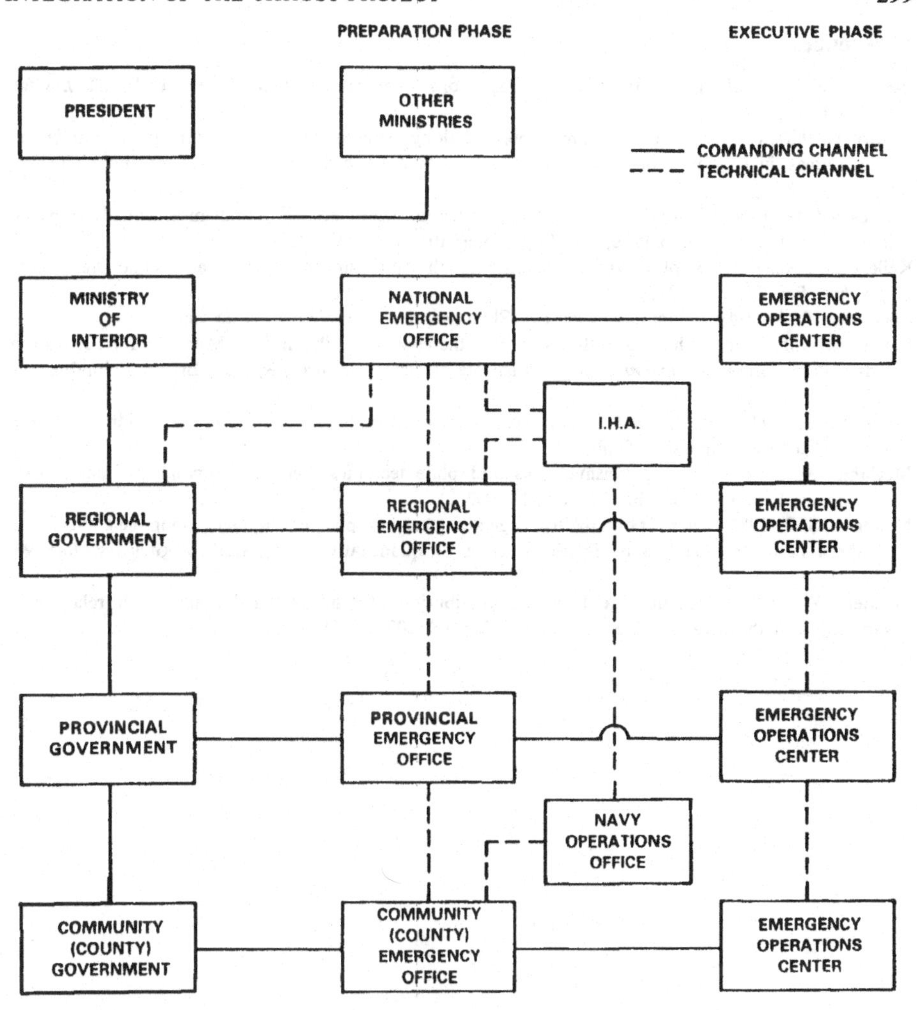

Fig. 4. National emergency organization chart.

- The Standard Operations Plan can be applied to any coastal community, providing that they have an emergency organization. It should be continuously tested to keep disaster planning and emergency personnel familiarized with the procedures. A twice-a-year simulation exercise is recommended.
- Utmost efforts should be taken to fund the improvement of seismic coverage and the seismic operations capabilities of the country.
- Reliable communications with PTWC are the key for a good operation of the National Tsunami Warning System. A backup of the commercial communications links is advisable.

References

Bernard, E. N. *et al.*: 1988, On mitigating rapid onset natural disasters: Project THRUST *Eos* **69**, 659–661.

Hatori, T.: 1983, Colombia-Perú tsunamis observed along the coast of Japan – Tsunami magnitude and source areas, in K. Iida *et al.* (eds), *Tsunamis – Their Science and Engineering*, Terra Pubs., Tokyo, pp. 173–183.

Hebenstreit, G. T. and Whitaker, R. E.: 1983, Tsunami hazard modeling and mitigation: Runup and inundation studies, SAI-83/1236, Sci. Appl. International Corp., McLean, Va.

Kelleher, J. A.: 1972, Rupture zones on large South American earthquakes and some predictions, *Geophys. Res.* **77**, 2077–2103.

Lockridge, P. A.: 1985, Tsunamis in Perú – Chile, Report SE-39, World Data Center A.

Lorca, E.: 1985, Tsunami hazard evaluation along the coast of Chile, in T. S. Murty and W. J. Rapatz (eds.), *Proc. Internal. Tsunami Sympos.*, Institute of Ocean Sciences, Sydney, British Columbia, pp. 193–199.

Lorca, E. 1987, THRUST project tsunami emergency operations plan for Chile, Inst. Hidrograf. de la Armada Report, Valparaíso, Chile.

McCann, W. R. *et al.*: 1978, Seismic gaps and plate tectonics: Seismic potential for major plate boundaries, U.S.G.S. Open File Report, 78–943.

Nishenko, S. P.: 1985, Seismic potential for large and great interplate earthquakes along the Chilean and southern Peruvians margins of South America: A quantitative reappraisal, *J. Geophys. Res.* **90**, 3589–3615.

Stauder, W.: 1973, Mechanism and spatial distribution of Chilean Earthquakes with relation to subduction of the oceanic plate. *J. Geophys. Res.* **73**, 5033–5061.

Natural Hazards **4**: 301–316, 1991.

The Tsunami Threat on the Mexican West Coast: A Historical Analysis and Recommendations for Hazard Mitigation

SALVADOR F. FARRERAS
División de Oceanología, Centro de Investigación Científica y de Educación Superior de Ensenada (CICESE), Apartado Postal 2732, Ensenada, Baja California 22800, Mexico

and

ANTONIO J. SANCHEZ
Estación de Investigación Oceanográfica, Secretaría de Marina, Vicente Guerrero 133 altos, Ensenada, Baja California 22830, Mexico

(Received: 10 January 1990; revised: 25 July 1990)

Abstract. From an inspection of all tide gauge records for the western coast of Mexico over the last 37 years, a data base of all recorded tsunamis was made. Information on relevant historical events dating back two centuries, using newspaper archives, previous catalogs, and local witness interviews, was added to produce a catalog of tsunamis for the western coast of Mexico. A description of the 1932 Cuyutlán tsunami is given. This is considered to be the most destructive local tsunami which has ever occurred in the region for which historical accounts are available. It was preceded by two precursor events, a not uncommon occurrence in that zone. A summary of the generation and coastal effects from the 1985 Michoacán tsunamis is also given. These Michoacán tsunamis are the most recent local events in that zone.

This information, and knowledge of local undersea faulting characteristics along the Mexican Pacific coast, leads to a clear differentiation of two zones of potential tsunami hazard: locally generated tsunamis south of the Rivera fracture, in the Cocos plate subsidence region, and remotely generated tsunamis north of this zone. Based on this zonation, two types of tsunami warning systems are proposed: real-time for the southern zone, and delayed-time for the northern. A description is provided of the Baja California Regional Tsunami Warning System that is presently operational in the northern zone.

Several major industrial ports and tourist resort areas are located in the southern zone, and are therefore most vulnerable to local destructive tsunamis. Some of these sites represent important socioeconomic resources for Mexico, and have therefore been chosen for a vulnerability assessment and microzonation risk analysis. Land use patterns are identified, risks defined, and recommendations to minimize future tsunami impact are given. One case is illustrated.

Key Words. Tsunami catalog, precursors, warning system, public education, risk assessment, Mexico, Mesoamerican subduction.

1. Introduction

The main goals of tsunami research include: helping to save lives, reducing property damage, and minimizing lifeline disruption and economic dislocation. These goals can be reached through a better understanding of the nature and extent of exposure to the phenomenon, the implementation of a reliable early warning system, and the development of enhanced pre-event preparedness programs. To achieve these goals in Mexico, the authors have started a preliminary action program with the cooperation of their institutions and other government agencies. This preliminary program is presently limited to the following elements:

(a) Compilation of an historic and recorded tsunami data base catalog for the western coast of Mexico.
(b) Implementation of the Baja California Regional Tsunami Warning System (BCRTWS); in the near future, the BCRTWS and a National Tsunami Warning System for Mexico will both be linked to the Pacific Tsunami Warning System (PTWS), with public education policies as a priority consideration.
(c) Microzonation of coastal areas according to an assessment of their vulnerability to tsunamis and their land-use patterns, followed by recommendations on countermeasures and an elaboration of the preparedness programs.

This approach may be considered as a prototype example of a disaster preparedness strategy that links scientific knowledge with socioeconomic policies, within the framework of the International Decade for Natural Disaster Reduction.

2. Data Base – Catalog and Precursor Events

Recognizing the importance of historical data bases for operational analysis, the Intergovernmental Oceanographic Commission (IOC) outlined a historical data and information program for tsunamis (IOC, 1987). Pararas-Carayannis (1987), as Director of the International Tsunami Information Center (ITIC), recommended that a comprehensive and exhaustive historic tsunami study be done in Mexico. Following the IOC and ITIC guidelines, a systematic compilation of tsunami information from newspaper archives, previous catalogs, historical reports, recent mareographic records, and local witness interviews was conducted. This information was structured and assembled in a bilingual (Spanish-English) catalog of tsunamis on the western coast of Mexico (Sánchez and Farreras, 1990).

A total of 68 records from 21 events (9 local and 12 remote sources) detected during the 37 years of operation of the tide gauge network since 1952, were found and reproduced in the catalog. Most of these rather recent events were small in nature and did not pose any substantial threat. Figure 1 summarizes date, gauge location, and source type of the records. Over the last 10 years, only the Acapulco station has produced usable tsunami records. All other tidal stations were either inoperable, or were converted to digital recorders with a 30 min sampling interval. Rehabilitation of some old stations is highly desirable to support present and future tsunami research in Mexico.

The catalog also includes detailed information on the first wave and maximum rise or fall parameters, descriptive accounts of the events from historical chronicles, and photographs. Contrary to what is generally believed, historic accounts from the

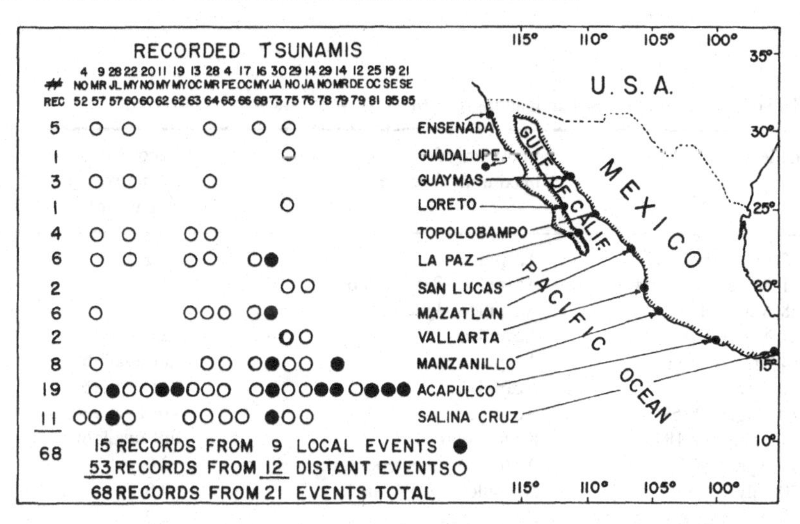

Fig. 1. Date, gauge location, source type, and number of tsunami records in existence, from the western coast of México.

last three centuries show that tsunamis that are locally generated in the Mesoamerican subduction zone constitute a serious threat to the southwestern coast of Mexico. The well-documented 16 November 1925 Zihuatanejo and 22 June 1932 Cuyutlán tsunamis, with estimated 7 to 11 and 10 m maximum wave heights respectively, were the two largest and most destructive contemporary events in the zone. Table I lists the date, the recorded or estimated maximum wave height, if available, and the location of primary observed damage or coastal effects for the destructive events. Table II lists the date, site, and height for the most recent non-destructive events.

The 10 m Cuyutlán tsunami of 22 June 1932 (7.7 Ms earthquake) was preceded by two smaller tsunamis on the 3rd and 18th of the same month with wave heights of 2 and 1 m (Excelsior, 1932a; El Universal 1932a), generated by two earthquakes with epicenters in the same region, but with magnitudes of 8.2 and 7.8 Ms, respectively (Geller and Kanamori, 1977). A similar sequence was also observed for the following strong seismic events:

(a) 7.6 Ms on 20 January and 7.1 Ms on 16 May 1900 in Jalisco;
(b) 7.5 Ms on 26 February, 7.8 Ms on 18 April 1902, and 7.7 Ms on 16 July 1903 in Chiapas;
(c) 8.1 Ms on 14 April 1907, 7.8 Ms on 29 March 1908, 7.5 Ms on 30 July 1909, and 7.8 Ms on 16 December 1911 in Guerrero;
(d) 7.7 Ms on 22 March, 8 Ms on 17 June, 7.6 Ms on 4 August, and 7.8 Ms on 9 October 1928 in Oaxaca;
(e) 6.6 Ms on 11 May and 6.5 Ms on 19 May 1962 in Guerrero; and
(f) 8.1 Ms on 19 September and 7.5 Ms on 21 September 1985 in Michoacán; as documented by Nishenko and Singh (1987).

Table I. Destructive tsunamis along the Mexican west coast

Date	Location of maximum damage	Recorded or estimated wave height (m)
25 February 1732	Acapulco	3.0
1 September 1754	Acapulco–San Marcos	4.0
28 March 1787	Acapulco–Igualapa	3.0
4 May 1820	Acapulco	2.0
14 March 1834	Acapulco	not available
7 April 1845	Acapulco	not available
12 August 1868	Acapulco	not available
24 February 1875	Manzanillo	not available
14 April 1907	Acapulco–Ometepec	2.0
30 July 1909	Acapulco–San Marcos	9.0
16 November 1925	Zihuatanejo	11.0
16 June 1928	Puerto Angel	6.0
22 June 1932	Cuyutlán–San Blas	10.0
28 July 1957	Acapulco	2.6
22 May 1960	Zihuatanejo	3.0
28 March 1964	Ensenada	2.4
29 November 1978	Salina Cruz-Puerto Escondido	1.5
19 September 1985	Lázaro Cárdenas	2.5

Table II. Non destructive tsunamis recorded at tidal stations along the Mexican west coast

Date	Location of tidal station	Maximum wave height (m)
4 November 1952	Salina Cruz	1.22
9 March 1957	Ensenada	1.04
20 November 1960	Acapulco	0.13
11 May 1962	Acapulco	0.81
19 May 1962	Acapulco	0.34
13 October 1963	Salina Cruz	0.49
4 February 1965	Salina Cruz	0.46
17 October 1966	Salina Cruz	0.24
16 May 1968	Acapulco	0.43
30 January 1973	Manzanillo	1.13
29 November 1975	Ensenada	0.46
14 January 1976	Manzanillo	0.21
14 March 1979	Acapulco	1.31
12 December 1979	Acapulco	0.30
25 October 1981	Acapulco	0.09
21 September 1985	Acapulco	1.20

The duration of these sequences is on a scale of days, months, or even a few years. Some sequences are composed of a main shock followed by aftershocks in a series of events that rupture the same segment of the plate interface (Anderson *et al.*, 1985), and others are complex modes of rupture of different segments of the plate boundary, which take several years to complete (Nishenko and Singh, 1987). In all cases, at least one and sometimes all of the seismic events in a sequence generated small or large tsunamis. This sequential pattern of strong tsunamigenic earthquake occurrence seems to be characteristic of the Mesoamerican subduction zone, and should be exploited to provide key precursor event information for incorporation into the tsunami warning policies of the region.

3. The Great June 1932 Cuyutlan Tsunami

At approximately 7 a.m. local time, on 22 June 1932, a strong earthquake (Ms = 7.7; Gutenberg and Richter, 1954) affected the coastal region of the States of Jalisco and Colima. No more than three minutes later, the ocean waters in front of Cuyutlán beach resort receded violently 300 to 400 m, piling up one over the other in a layer-like manner until it reached the appearance of a huge vertical wall about 9 to 10 m high (Excelsior, 1932b). Suddenly, with unexpected speed, the water advanced toward the beach, preserving the appearance of a vertical wall, until it reached the two-story Santa Cruz Hotel building, which was situated on a sand bar that separated the village from the sea. After overtopping the building, the huge wave broke onto the village square, lifting up stone paving blocks on the main street, eroding a trench, and falling like a 2–3 m thick blanket that flooded the entire village (recounted by witnesses Pio Ventura and Heriberto Sánchez). Everything was washed away and destroyed, included five large hotels, the church, about 80 wooden houses, and several small cottages; even the Hotel Ceballos, the only concrete structure in town, could not withstand the tsunami (Excelsior, 1932c). A truck loaded with sand was carried 200 m inland. The wooden floor of the church was transported 500 m. The water invaded a distance of about 1 km inland within 2 to 3 min, until it reached the railroad tracks embankment, where it deposited trees, sand and debris, obstructing a 2 km long extension of track (El Nacional, 1932a). A few people and horses were left at the top of palm trees. It took the water approximately two hours to return to its normal state (El Universal, 1932b). After the tsunami waves receded, the village was strewn with a mixture of human corpses, dead sharks and fish, cattle, trees, sand and debris. Ten to seventy-five casualties were estimated by different sources. The waves flooded destructively alongshore for approximately 5 km, but the phenomenon was observed for 25 km along the coast (El Nacional, 1932b). Other smaller coastal communities nearby, such as Tenancillo and Palos Verdes, were completely destroyed (El Universal, 1932c).

Two additional tsunamis which were smaller and reached only the main square of Cuyutlán, were caused by aftershocks of the main earthquake over the next two days (El Universal, 1932b).

4. The September 1985 Michoacán Tsunamis

On 19 September 1985 an 8.1 Ms earthquake, composed of two subevents with 80 km source separation and a 26 sec time lag, occurred in the subduction zone of the northwest portion of the Cocos Plate (Anderson *et al.*, 1986). The generated tsunami affected several coastal communities in the States of Michoacán and Guerrero. Waves of approximately 2.5 m height arrived at the heavy industry-based port of Lázaro Cárdenas during the earthquake shaking, and flooded a horizontal distance of approximately 500 m inland. The estimated runup was based on the finding that the waves covered a dock that stands 3.15 m above mean sea level. Jetties at the entrance of the harbor were completely covered by the waves, but were left undamaged. Major damage was attributable to inundation. Approximately 1.5 km of railroad track were destroyed because the bank under the track was eroded away. The fence protecting the railroad tracks from storm and tidal action was washed out. An earth embankment bridge providing access to a fertilizer plant was also washed out. Little or no impact or drag effect were observed; the water flow neither tipped over empty railroad cars on the tracks close to the shore nor removed the rocks on the beach groins and river entrance jetties. Beach configuration changes attributable to approximately 2 m vertical piling of sand were also observed. Results of a survey with detailed damage patterns are presented elsewhere by Urban Regional Research (1988). A large perturbation followed by a train of small waves, resembling an undular bore was observed propagating upstream as far as the second bridge, 7.5 km from the river mouth (Farreras and Sánchez, 1987).

In Playa Azul, a local tourism-based coastal community, there were four to five 2.5 m high waves filling and emptying the beach approximately once every half-hour. The tsunami waves invaded a distance of about 100–150 m inland, carrying away the frail palapa-built restaurants and furniture and depositing a ledge of sand approximately 25 m wide × 2 m high at the upland edge of the beach.

In Ixtapa, an international tourism-based coastal community, four to five tsunami waves were observed, and it was reported that the water receded, then the bay filled up and the beach became 'like a bathtub' several times. At the highest runup the water came over a 1.5 m wall and into the swimming pool of the Sheraton Hotel.

On 21 September 1985, a 7.5 Ms aftershock occurred in the southern portion of the so-called 'Michoacán seismic gap', with a much smaller rupture zone area than the previous main shock. The generated tsunami affected mainly the traditional fishing village of Zihuatanejo. The water in the bay receded, then there were seven to eight 2 m to 3 m high waves, and then the water remained oscillating for 9 h due to the resonant conditions of the semi-enclosed bay. The water overtopped the Municipal Pier and invaded a distance of about 200 m inland. Considerable flood damage was caused by the water that came to the first floor of the major beachside hotels and restaurants. A similar oscillating resonant pattern was observed in Zihuatanejo during the devastating 11 m wave height tsunami on 16 November 1925.

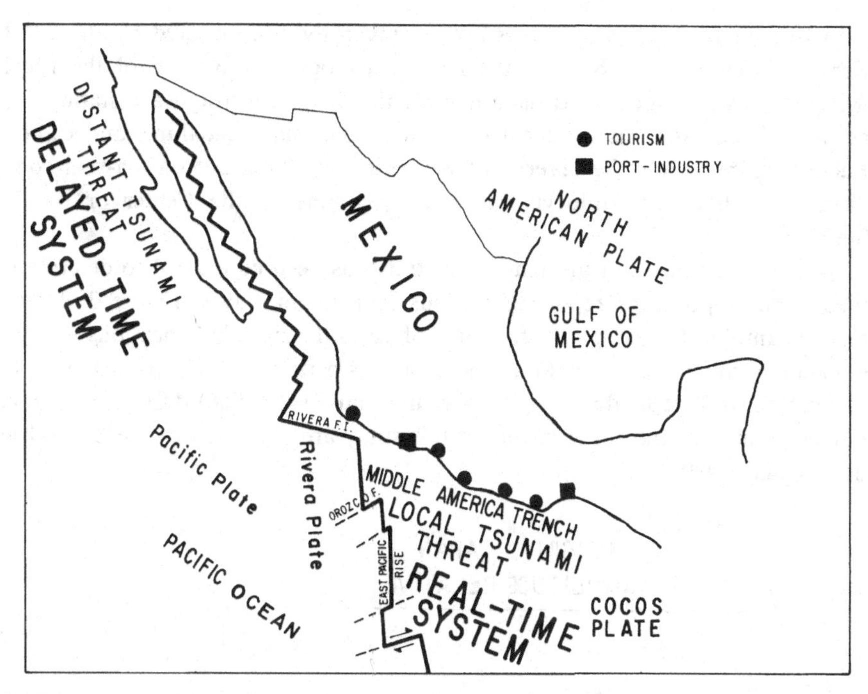

Fig. 2. Seismotectonic setting, predominance of tsunamis, and type of warning systems proposed along the Mexican Pacific coast.

No casualties were reported in any of the affected communities as a direct result of the 19 and 21 September 1985 tsunamis.

5. Tsunami Warning Systems

Based on the nature of undersea faulting and tectonic plate interaction, Sánchez and Farreras (1987) defined two hazardous zones along the Mexican west coast, each with different implications for tsunami generation and arrival (Figure 2). Southeast of the East Pacific Rise and along the Middle America Trench, the Cocos Plate subsides beneath the North American Plate at rates of 2.5 to 7.7 cm/yr (McNally and Minster, 1981). Large earthquakes occur at fairly regular intervals in this zone, and some have produced devastating local tsunamis. Northwest of the East Pacific Rise, the Pacific Plate slides north with respect to the North American Plate, along the Gulf of California strike-slip fault; and the Rivera Plate rotates without subduction at a rate of 1.6 cm/yr at the margin. Generation of large tsunamis in this zone is unlikely, as confirmed by historical information; but several small and moderate tsunamis generated by remote sources, have been recorded.

The distinctive nature of the tsunami threat in each zone indicates the need for two different types of warning system: a real-time system in the southern zone, and a delayed-time system in the northern zone.

In the northern zone, the BCRTWS is presently administered by the CICESE Research Center. CICESE and the local Oceanographic Station of the Mexican Secretary of the Navy in Ensenada operate the three reporting wave stations in the network. These stations are located in Ensenada, Isla Guadalupe, and Cabo San Lucas (Figure 1), and the observers have permanent 24 hr/365 days-a-year communication facilities available via radio or telephone to the Center or the Navy Station.

To improve siting and installation, a study has been conducted to determine the theoretical response of reporting stations to tsunami waves. Figure 3 shows the relative amplitude and phase response of several azimuthal locations along Isla Guadalupe contour to tsunami waves of various periods arriving from Japan. These computations were made using the Vastano and Reid (1966) diffraction-refraction numerical model, and other results of this study are presented by Reyes-Rodríguez de la Gala (1990).

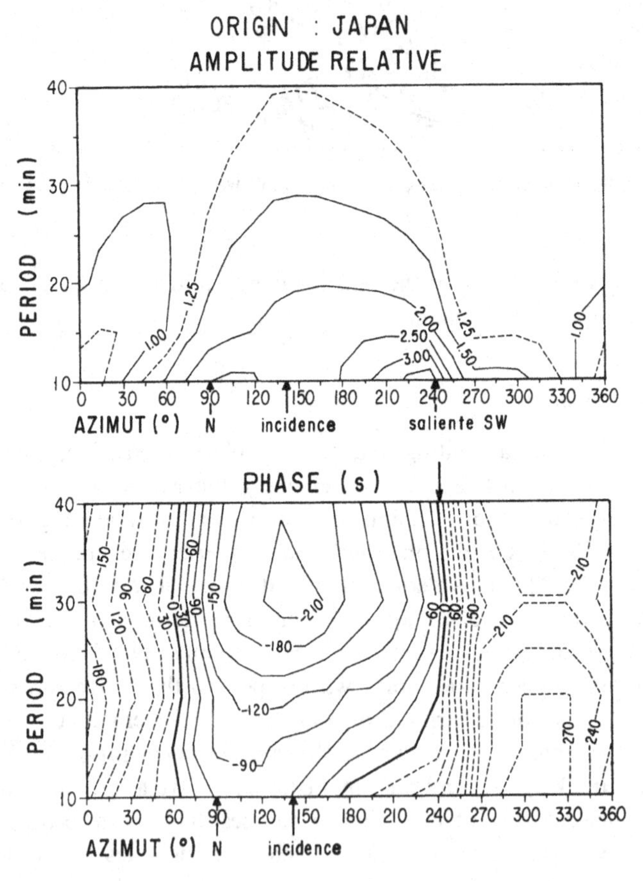

Fig. 3. Relative amplitude and phase response of azimuthal locations around Isla Guadalupe to tsunami waves incident from Japan.

The CICESE Research Center also acts as a dissemination agency for messages to and from the Pacific Tsunami Warning Center. Cabo San Lucas presently operates as a wave reporting station for the PTWS (National Oceanic and Atmospheric Administration, 1988), and the addition of the Isla Guadalupe station to the network has been requested by the IOC (1987).

The BCRTWS works jointly with the Baja California Civil Protection System (BCCPS), and the local army and navy authorities. Figure 4 shows the operational diagram of the BCRTWS. Ten thousand copies of a brochure in Spanish titled 'Que Hacer en Caso de Maremoto' (What to do in a Tsunami event) were prepared by the authors under the framework of the BCRTWS; these were edited and then distributed by the BCCPS among coastal community residents as part of a public education program for disaster prevention (Figure 5). Tsunami drills have been conducted by BCCPS with the cooperation of BCRTWS. These two agencies have also recently initiated a joint study of land-use patterns, evacuation routes, and shelter and services distribution for urban planning along coastal zones in Baja California that are vulnerable to tsunamis.

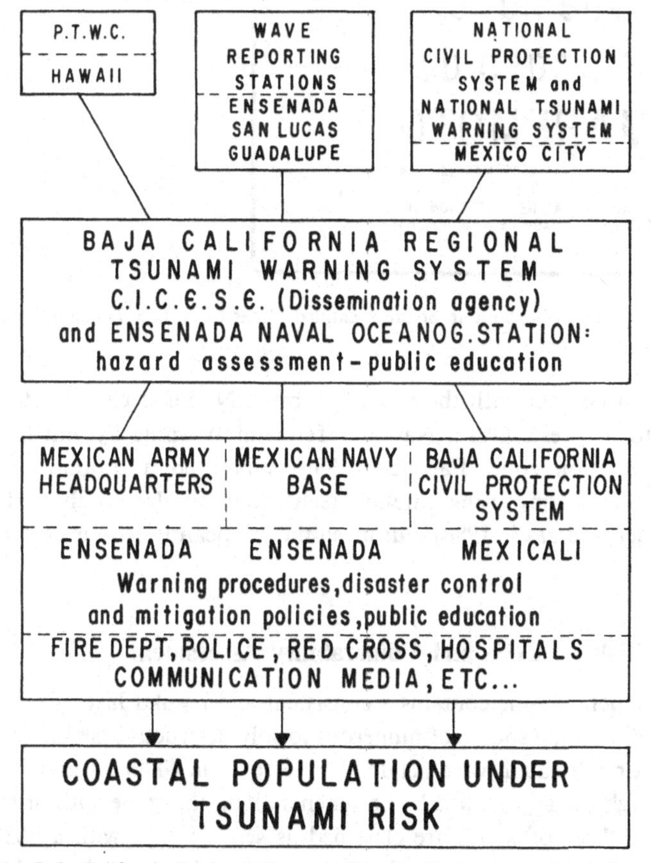

Fig. 4. Baja California Regional Tsunami Warning System operational diagram.

Fig. 5. Tsunami public education brochure in Spanish, entitled 'What to do in a Tsunami event'.

BCRTWS is also in contact with the recently-formed National Civil Protection System of Mexico to help establish a National Tsunami Warning System for the southern active subduction zone. Farreras (1986) submitted a proposal for a real-time system for this zone using satellite technology similar to that of the THRUST project (Bernard *et al.*, 1988), but using the Mexican telecommunications satellite 'Morelos'.

6. Microzonation Risk Analysis and Vulnerability Assessment

The Cocos Plate subduction zone contains a consistent and regular history of great earthquakes, and has been the source of numerous locally destructive tsunamis with maximum recorded or estimated wave heights of 1.5 to 11 m (Table I and Figure 6). Nishenko and Singh (1987) identified 13 segments along this zone with histories of large earthquakes, three of which are classified as seismic gaps with a 60% to 80% probability over the next 10 years of the recurrence of magnitude 7.5 Ms or

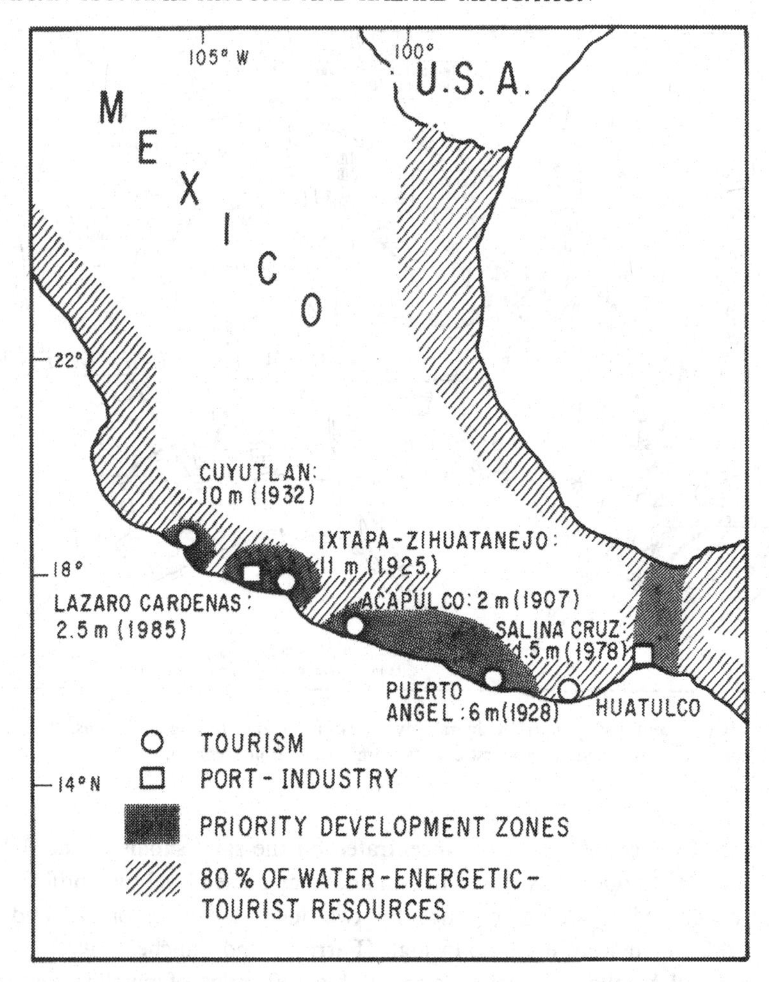

Fig. 6. Main industrial ports and tourist resorts along the priority development zone identified by the National Urban Development Plan of Mexico, with maximum wave heights and dates of occurrence of past tsunamis.

larger events. The highly vulnerable coastal area adjacent to this subsidence zone is also considered to be a priority area for decentralization and development by the National Urban Development Plan of Mexico. It possesses an important part of the water-energy-tourism resources of the nation, and is the site of two major industrial ports: Lázaro Cárdenas and Salina Cruz. Approximately 1000 km of tourist resorts form a corridor from Cuyutlán to Huatulco (Figure 6).

Facilities in Lázaro Cárdenas include two steel mill plants, a fertilizer factory, a container port, metal and mineral docks, grain storage silos, and a 45 ha marine terminal under construction. The marine terminal will be capable of storing 541 000 million barrels of gasoline, diesel, jet fuel, and fuel oil. The estimated population is

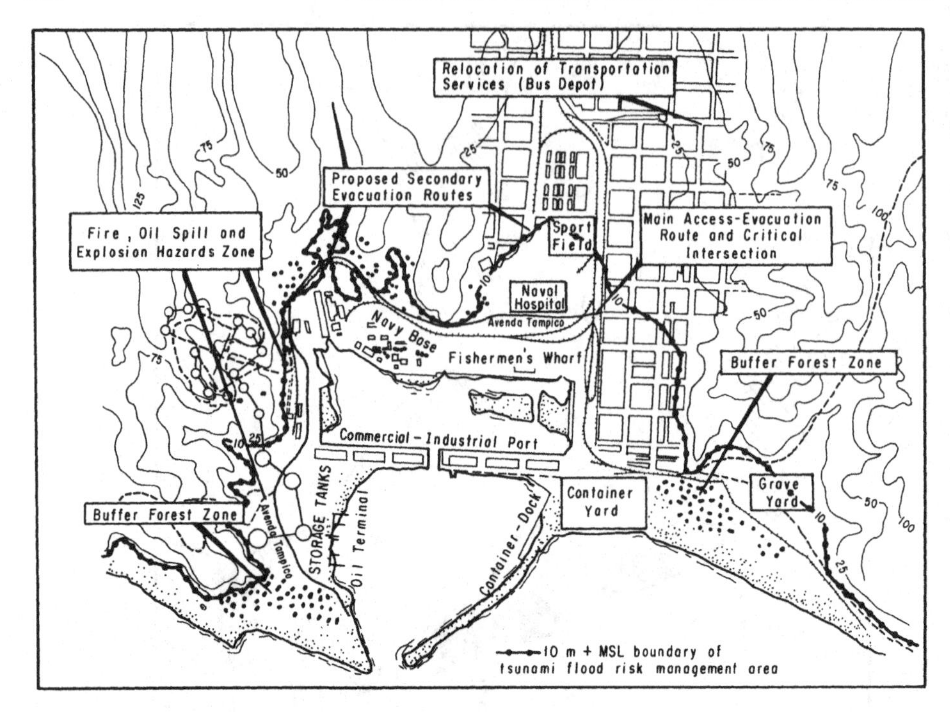

Fig. 7. Salina Cruz land use pattern, vulnerability assessment, and recommendations. The 10 m flooding zone contour corresponds to a worst-case tsunami inundation scenario.

150 000. Most of the construction is concentrated on the sand shoals of the Balsas River delta, after the region was raised to 5 m above mean sea level by landfill. This area recently suffered moderate damage by two local tsunamis on 19 and 21 September 1985 which were each 2.5 m high (Farreras and Sánchez, 1987).

Salina Cruz will be the west coast terminus for 160 miles of pipeline that will carry crude oil across the Tehuantepec Isthmus to the Pacific, alleviating the need for tankers to travel the Panama Canal. This port presently has a container terminal, facilities for bulk cargo, fuel storage tanks and a 400 vessel fishing fleet (Figure 7). A 150 000 barrels-per-day expansion of the present oil refinery and export terminal will be completed in the near future.

Typical of the many new tourist resorts now under development are the 2 500 ha of beaches of the nine Huatulco Bays. These lie along an 18 mile stretch north of Salina Cruz, and are separated from each other by tropical vegetation and steep hillsides. The National Tourism Development Bureau (FONATUR) has invested over 10 million dollars in infrastructure to support the proposed development; water and sewer facilities, a coastal highway and an international airport have been completed to date. A Sheraton and a Club Med Hotel have been built on the beaches at Tangolunda Bay, and FONATUR plans that hotel rooms will increase from 850 in 1989 to 6700 by 2018, with a corresponding population increase from

12 800 to 308 000. Most of this tourist development is in the coastal zone and are thus vulnerable to tsunami wave impact.

Several of these locations were selected for a microzonation risk analysis and vulnerability assessment study; such studies are aimed at reducing the loss of life and property, and minimizing the socioeconomic disruption caused by tsunamis.

The methodology to use in a microzonation risk analysis includes the following elements:

(a) Identification of a target area for the study according to historical evidence of past tsunamis, neighbouring tectonic scenarios such as seismic gaps, present and future socioeconomic development, and suspected implications of a disruption by a major tsunami.

(b) Determination of probable tsunami wave elevations and expected inundation limits, through some of the following alternatives: calculation of wave-height probabilities from tidal harmonic constituents and recorded or estimated historic tsunami heights, wave-diffraction diagrams from rupture zones that generated tsunamis in the past or are suspicious to generate in the future, direct assignment of the calculated shoreline runup heights to the entire reach of the inland zone, and numerical simulation of the wave propagation from the source to the shoreline and inland areas considering, among other factors, the beach slope, bottom friction, tectonic subsidence or uplifting, wave profile and directionality, and the presence of vegetation and man-made structures. The boundaries of the inundation area will define a Flood Damage Control Zone.

(c) Field survey within the Flood Damage Control Zone to establish with the help of aerial photographs, municipal charts and local urban development plans, the present and future trends in land use patterns, including: population density distribution, location and type of public office buildings, schools, hospitals, hotels, industrial installations, port facilities, marine structures, businesses, family housing, public services, utility plants and networks, fuel storage tanks, resort areas, vacant lots, roadways and transportation networks. Information on soil stability should be added too.

(d) Assessment of vulnerability to primary causes (inundation, hydrodynamic forces and loss of ground support) and secondary causes (impact of floating debris, hazardous releases, fire and contamination) of the population, buildings, marine structures, lifelines, and others (beach facilities, aquaculture).

(e) Mapping of risk zones based on the land use patterns and the vulnerability assessment.

(f) Development of recommendations for countermeasures such as the redistribution of population, structures and services, improvement of evacuation routes, location of shelters and buffers that may act as barriers against the tsunami action and implementation of a public education program.

Some of the guidelines given above and their application to the microzonation risk analysis of Ixtapa-Zihuatanejo and Lázaro Cárdenas are presented elsewhere (Urban Regional Research, 1988).

Land-use patterns, vulnerability assessment, and recommended countermeasures for Salina Cruz are summarized in Figure 7 (Preuss *et al.*, 1989). The contour line corresponding to 10 m above the mean sea level delineates the worst case scenario of maximum probable tsunami inundation, based on historical information along the entire subduction zone. Salina Cruz is characterized by intensive industrial use of the coastal areas and high day-time population density. Furthermore, the only access at present into and out of the port is an extremely congested intersection crossed by railroad tracks. The population is therefore left with almost no possibility of safe evacuation in the event of a tsunami. Recommendations for Salina Cruz include the establishment of alternative evacuation routes, buffer tree forestation near the beaches as an effective means of reducing tsunami wave impact on the container yard and downtown area, protection of the nearshore fuel storage tanks against tsunami wave or floating debris impact, the development of evacuation plans and drill exercises by the local authorities and the population, and implementation of a public education program on tsunami hazard mitigation policies.

7. Conclusions and Recommendations

Locally-generated tsunamis constitute a significant hazard to the southwest coast of Mexico along the Middle America Trench. A real-time tsunami warning system using satellite communications technology, should be implemented as soon as possible for this coastal area.

Strong tsunamigenic earthquakes in the subsidence zone have historically been characterized by a sequential pattern of occurrence that could be exploited to develop criteria for the identification of potential precursors to large tsunamis. This information could then be incorporated into the tsunami warning policies for the region.

The microzonation risk analysis and vulnerability assessment studies must continue and expand to cover all of the west coast communities of Mexico vulnerable to tsunamis. Numerical modeling should be conducted to provide estimates of inundation patterns for each site.

Public education on tsunami hazard mitigation must be included in disaster preparedness programs for each region.

Acknowledgements

Consejo Nacional de Ciencia y Tecnología (CONACyT) from México provided funds through Research Grant P218CCOC880065, and financed travel expenses through Special Support Grant ICCNXNA-031341.

Jane Preuss of Urban Regional Research performed important parts of the microzonation risk analysis and vulnerability assessment study under National Science Foundation Grant INT-8709972.

This paper was presented at the IUGG International Tsunami Symposium held in Novosibirsk, U.S.S.R., August 1989.

References

Anderson, J. G., Bodin, P., Brune, J. N., Prince, J., Singh, S. K., Quaas, R., and Onate, M.: 1986, Strong ground motion from the Michoacán-México earthquake, *Science* **233** (4768), 1043–1049.

Bernard, E. N., Behn, R. R., Hebenstreit, G. T., González, F. I., Krumpe, P., Lander, J. F., Lorca, E., Mc Manamon, P. M., and Milburn, H. B.: 1988, On mitigating rapid onset natural disasters: Project THRUST, *EOS Trans. Amer Geophys. Union* **69** (24), 649–661.

El Nacional: 1932a, La población de Cuyutlán fué destruida por gigantesca ola, *Diario Popular*, México DF, Jueves 23 de Junio, **XIII** (1122), 1.

El Nacional: 1932b, Todo cuanto encontró a su paso arrasó furiosamente el mar, *Diario Popular*, México DF, Viernes 24 de Junio, **XIII** (1123), 1.

El Universal: 1932a, El mar subió casi dos metros, *El Gran Diario de México*, México DF, Domingo 5 de Junio, **LXII** (5695), 1–7.

El Universal, 1932b, Dos veces volvió el Pacífico a lanzar sus aguas sobre Cuyutlán, *El Gran Diario de México*, México DF, Viernes 24 de Junio, **LXII** (5714), 1.

El Universal: 1932c, Una ola gigantesca arrasó ayer Cuyutlán; hubo 30 muertos, *El Gran Diario de México*, Mexico DF, Jueves 23 de Junio, **LXII** (5713), 1–5.

Excelsior: 1932a, Personas que llegaron hoy de Manzanillo informan, *El Periódico de la Vida Nacional*, México DF, Lunes 20 de Junio, **III** (5556), 1.

Excelsior: 1932b, Excelsior en el lugar de la espantosa catástrofe, *El Periódico de la Vida Nacional*, México DF, Sábado 25 de Junio, **III** (5561), 1–3.

Excelsior: 1932c, Dos veces mas Cuyutlán ha sido barrido por las olas, *El Periódico de la Vida Nacional*, México DF, Viernes 24 de Junio, **III** (5560), 1–3.

Farreras, S. F.: 1986, *Sistema Regional de Alarma de Maremotos (Tsunamis) para Baja California*, Misc. Tec. Rep. a la Subdirección Nacional de Explotación de Satélites SCT, Centro de Inv. CICESE, Ensenada, BC.

Farreras, S. F. and Sánchez, A. J.: 1987, Generation, wave form and local impact of the September 19, 1985 Mexican tsunami, *Sci. Tsunami Hazards* **5**, 3–13.

Geller, R. J. and Kanamori, H.: 1977, Magnitudes of grat shallow earthquakes from 1904 to 1952, *Bull. Seismol. Soc. Am.* **67**, 587–598.

Gutenberg, B. and Richter, C. F.: 1954, *Seismicity of the Earth and Associated Phenomena*, Princeton Univ. Press, Princeton, New Jersey.

Intergovermental Oceanographic Commission: 1987, *Master Plan for the Tsunami Warning System in the Pacific*, 10C/INF-730, UNESCO, Paris.

McNally, K. C. and Minster, J. B.: 1981, Non uniform seismic slip rates along the Middle America Trench, *J. Geophys. Res.* **86**, 4949–4959.

National Oceanic and Atmospheric Administration: 1988, *Communication Plan for the Pacific Tsunami Warning System*, R. L. Sillcox and W. M. Mazey (compilers), Nat. Weather Serv., U.S. Dept. of Comm., Ewa Beach, Hawaii.

Nishenko, S. P. and Singh, S. K.: 1987, Conditional probabilities for the recurrence of large and great interplate earthquakes along the Mexican subduction zone, *Bull. Seismol. Soc. Am.* **77**,

Pararas-Carayannis, G.: 1987, *International Tsunami Information Center a Progress Report for 1985–1987*, XI Session of the ICG for the PTWS, Beijing, China.

Preuss, J., Farreras, S. F. and Sánchez, A. J.: 1989, *Planning for Seismic Hazards Microzonation of Southwest Mexico Coastal Areas*, Urban Regional Research, Seattle, Washington.

Reyes-Rodríguez de la Gala, J.: 1990, *Respuesta Lineal en Amplitud y Fase de Isla Guadalupe* (*Baja California, México*) *a Ondas de Tsunamis*, Tesis de Oceanólogo, Univ. Aut. de Baja Cal., Ensenada, BC.

Sánchez, A. J. and Farreras, S. F.: 1987, Tsunami threat to the Mexican Pacific ocean coast-summary, in E. N. Bernard and R. L. Whitney (eds.), *Proceedings of the International Tsunami Symposium, IUGG. 18 to 19 Aug 1987*, NOAA/ Pac. Marine Env. Lab., Seattle, Washington, pp. 215–219.

Sánchez, A. J. and Farreras, S. F.: 1990, *Catálogo de Maremotos (Tsunamis) en la Costa Occidental de México*, Centro de Inv. CICESE y Sec. de Marina de México, Ensenada, BC.

Urban Regional Research: 1988, *Planning for Risk: Comprehensive Planning for Tsunami Hazard Areas*, prepared for The National Science Foundation, Seattle, Washington.

Vastano, A. C. and Reid, R. O.: 1966, *A Numerical Study of the Tsunami Response at an Island*, Ref. 66-26T, Dept. of Oceanography, Texas A&M University, College Station, Texas.

Natural Hazards 4: 317–326, 1991.

Meeting Reports

Twelfth Session of the International Coordinating Group, Tsunami Warning System in the Pacific (ICG/ITSU), Novosibirsk, U.S.S.R., 7–10 August 1989

In 1965, the Intergovernmental Oceanographic Commission (IOC) accepted the offer of the United States to undertake the expansion of its existing Tsunami Warning Center in Honolulu to become the headquarters of the International Tsunami Warning System (ITWS). The IOC also accepted the offer of other Member States to integrate their existing facilities and communications into this International Warning System. At a meeting in Honolulu in 1965, an agreement was reached and IOC established the International Coordination Group for the Tsunami Warning System in the Pacific (ICG/ITSU).

The ICG/ITSU was established as a subsidiary body of IOC, meeting every two years to coordinate and review the activities of the International Tsunami Warning System. Twenty-four nations are now members of ICG/ITSU. Several nonmember states maintain stations. The System makes use of approximately 31 seismic stations, 53 tidal stations, and 101 dissemination points scattered across the Pacific under the varying control of the Member States of ITSU.

The role of the ICG/ITSU is to reduce the risk to lives and property in Member States whose coastal areas are threatened by tsunamis, and to carry out this role by recommending improvements to the TWS; by promoting regional cooperation between Member States; by contributing to the scientific and technical training of tsunami experts, and the education of the general public in tsunami awareness; by encouraging the development of improved instrumentation and communication systems; by ensuring the exchange of information between participating countries and between such organizations as the WMO and IUGG; and by offering assistance to the national and regional needs of Member States.

The Twelfth Session of the International Coordinating Group, Tsunami Warning System in the Pacific (ICG/ITSU), was held 7–10 August 1989 in Novosibirsk (Akademgorodok), U.S.S.R. The Session of the Group was the third of a series of meetings held in Novosibirsk from 31 July–10 August 1989. The first was the Symposium of the International Union of Geodesy and Geophysics Tsunami Commission (IUGG/TC). This was followed by a Workshop on the Technical Aspects of Tsunami Warning Systems, Tsunami Analysis, Preparedness, Observation, and Instrumentation. This was jointly sponsored by the IUGG/TC and the ICG/ITSU.

The excellent facilities of the Siberian Branch of the U.S.S.R. Academy of Sciences were provided for all meetings. The support and other amenities provided by the Local Organizing Committee resulted in a most productive and enjoyable experience. A total of 12 Member States were present for the Session. Except for ITSU-X at Sydney, BC, this was the greatest number of Member States in attendance since the initial Session in 1965. A Summary Report of the Session is available from the IOC, Paris, as Document IOC/ITSU-XII/3.

While the usual amount of routine business was conducted during the Session, it was quite obvious that the Group's main focus was on the contributions which it might make in furthering the aims of the International Decade on Natural Disaster Reduction (IDNDR). The Group is in an ideal position to act in this area. It not only has many years of experience in operating a real-time system which is directed at natural disaster reduction, but the Group has a continuing and demonstrated interest in working with developing countries.

During the Session, the Group learned that signficant progress is being made toward obtaining funding for the establishment of the Group's long-desired South-west Pacific Sub Regional Tsunami Warning System. This System, with its associated tidal and seismic networks, will be of great benefit to all countries of the area. Success in obtaining the funding will bring to fruition a project which the Group began in 1982. (The five-year plan required by the UNDP has been submitted and a final decision on funding is pending.)

As a result of decisions at the Session, three working groups are being established to develop projects in direct support of the IDNDR. These are

As a result of a Chilean proposal, the Group recommended and the Chairman established, an *ad hoc* working group to develop a programme "... in the area of public awareness and education of tsunamis ..." A firm proposal is to be ready for consideration at ITSU-XIII. Canada, Chile, New Zealand, U.S.A., and the Director, International Tsunami Information Centre (ITIC), are represented on the working group. It is already at work collecting data.

Included in the recommendations of the workshop, and adopted by the Group, was one calling for the establishment of a joint IUGG/TC-ICG/ITSU *ad hoc* working group to formulate a joint project which is supportive of the efforts of the IDNDR. The project is to be ready for consideration and approval at ITSU-XIII. This working group has been formed and held its first meeting in Honolulu from 30 January–1 February 1990. A Report of actions was prepared and distributed for information and comment.

At the request of the Group, and responding to a proposal of the U.S.S.R., the Chairman and the IOC Secretariat agreed to the desirability of the establishment of a group of experts to look at the possibility of establishing real-time communications and computer interconnections between the U.S.S.R. and U.S.A. centres and possibly other centres. Letters have been sent to France, Japan, U.S.A., and the U.S.S.R. requesting them to nominate a member to serve on the group of experts.

Action was also completed on a number of long-standing items. For example:

Arrangements were completed for the distribution of the Atlas of Travel Time Charts prepared by the Computing Centre, Krasnoyarsk.

The contents of the first edition of the Glossary of Tsunami Related Terms was fixed and cleared for publication.

The Group approved the distribution of the Standardized Tsunami Database Format, data, and associated software.

The Group also adopted a recommendation of the workshop which recommended establishment of formal liaison between the ICG/ITSU and the Federation of Digital Broadband Seismograph Networks (FDBSN) and their close cooperation when establishing seismic stations for tsunami purposes.

The work of Dr Jacques Talandier, Directeur, Laboratoire de Géophysique, Tahiti, was recognized by the Group which encouraged him to continue his research on mantle magnitude. The Group also requested the operational centres in the System to incorporate the determination and use of mantle magnitude in their programmes.

The fact that tsunamis are not unique to the Pacific Basin was a subject of discussion at ITSU-XI and was considered again at this Session. While it was recognized that an ITSU-like function would be valuable in other areas, it was the consensus of the Group that an expansion of the ICG-ITSU mandate to include other areas would dilute significantly its effectiveness in the Pacific Basin. The Group believed, therefore, that it could make a more effective contribution by serving as a resource for similar groups which might be formed in other areas of the world.

Finally, the Group adopted the 'Work Programme and Priorities for 1990–1991'. This was a considerably more ambitious plan than adopted at previous sessions and will require a substantial increase in funding over the current biennium.

RICHARD H. HAGEMEYER
Chairman, ICG/ITSU

Second International IOC Workshop on the Technical Aspects of Tsunami Warning Systems, Tsunami Analysis, Preparedness, Observation, and Instrumentation, Novosibirsk, U.S.S.R., 4–5 August 1989

Background. During the past sessions of the International Co-ordination Group for the Tsunami Warning System in the Pacific (ICG/ITSU), great emphasis was placed on the educational program on tsunamis and the training of officials of ICG/ITSU member countries. This need was further emphasized by the United Nations' declaration of the next decade as the International Decade of Natural Disaster Reduction (IDNDR).

Tsunami is one of the major disasters that threatens the coastal populations of the world oceans and inland seas. The Tsunami Warning System in the Pacific (TWS) has been a major effort spearheaded by the Intergovernmental Oceanographic Commission (IOC) and its International Co-ordination Group for the Tsunami Warning System in the Pacific to mitigate the effects of the tsunami disaster. The Tsunami Warning System in the Pacific has been in existence since 1965. However, a great deal of progress has been made in the last few years on instrumentation, communications, and computer applications, which have had, or could have, great impact on the improvement of the Tsunami Warning System. The state of the art is rapidly changing. Even experts in the field have to review from time to time progress that is being made in technology to familiarize themselves with new concepts and learn to apply these concepts into operational techniques that can result in better tsunami analysis, prediction, and communications. Improvements can be obtained in data collection and rapid processing of data as well as in prediction of tsunami heights and inundation by applying the new technology and new instrumentation to gather, process, and analyze data. Therefore, a real need was identified to have workshops and training sessions, even for the experts, during which instruction and information can be given on new technological advancements, information concerning computer circuitry, data transmission techniques, data collection, and calibration techniques and communications. Training of officials involved in the Tsunami Warning System is an important part of the overall educational requirements of ITSU member countries because these officials are, in turn, responsible for operational improvements in their own countries and for a program of general public education.

As early as August 1983, the IOC Secretariat called a special meeting in Paris, which included the Chairman of ICG/ITSU and the Director of the International Tsunami Information Center (ITIC), to review the educational needs of ITSU members. Suggestions were made that tsunami workshops should be held under the auspices of the TEMA program, that a plan for a workshop should be drafted, and that appropriate experts should be designated for such training. ITIC was charged with the responsibility of developing a curriculum and locating instructors. It was also suggested that such a workshop could be held consecutively to the ITSU and IUGG sessions so as to maximize participation and minimize cost.

On the basis of these suggestions, ITIC, in close consultation with the IOC Secretariat and the Chairman of ICG/ITSU, developed a curriculum for the training of such officials and for familiarization of participants in the TWS, not only with conceptual improvements that have been made, but with the inner workings of the TWS, including computer applications, on-line processing, and numerical modeling. Thus, the first IOC-sponsored Workshop on the Technical Aspects of Tsunami Analyses, Prediction and Communications, was held at Sidney, BC, Canada, on 29–31 July 1985, prior to the ITSU-X Meeting, and prior to the IUGG Conference in nearby Victoria.

Four years have since elapsed, and in this time interval, the technology has greatly changed. This second Workshop on the Technical Aspects of Tsunami Warning Systems, Tsunami Analysis, Preparedness, Observation, and Instrumentation was held on 4–5 August 1989 at Novosibirsk, U.S.S.R. The purpose of this second Workshop was to bridge the gap of four years of independent developments in the TWS and to bring together tsunami specialists from different countries to improve their knowledge of the tsunami phenomenon and to help find practical solutions to the improvement of TWS for the mitigation of the tsunami hazard. As with the first Workshop, the second Workshop was held right after the IUGG Tsunami Conference and just prior to the XII Session of ICG/ITSU. The Workshop was part of the overall TSUNAMI 89 Conference and was held in the modern and attractive research town of Academgorodok, which is located 20 km south from downtown Novosibirsk, the capital of Siberia, in a pine-tree forest growing along the bank of the Ob River. The U.S.S.R. Academy of Sciences and the Computing Center of its Siberian Division hosted the TSUNAMI 89 Conference and this second Workshop.

A report containing the summary of the proceedings of this Workshop, as well as Annexes containing the Workshop Program, the Recommendations and a List of Participants will be published by UNESCO-IOC in the near future in its Workshop Report Series. The full text of papers presented at this Workshop will be published by UNESCO-IOC as a Supplement to the Summary Workshop Report.

The reader is referred to these UNESCO-IOC reports for detailed accounts of the Workshop proceedings. The following summary provides only the highlights and major conclusions and recommendations of the Workshop.

Workshop Summary. The Workshop on the Technical Aspects of Tsunami Warning Systems, Tsunami Analysis, Preparedness, Observation, and Instrumentation was opened at the Dom Uchenyh Hall at the scientific town of Akademgorodok, Novosibirsk, U.S.S.R. on 4 August 1989, at 9:00 am.

Dr Kazuhiro Kitazawa, Assistant Secretary of the Intergovernmental Oceanographic Commission (IOC) opened the Workshop, and speaking on behalf of the IOC Secretariat, extended to the participants of the Workshop a very warm welcome. Next, Mr Hagemeyer, Chairman of the International Coordination Group for the Tsunami Warning System in the Pacific (ICG/ITSU), speaking on behalf of the Member States of the Group, extended to the participants a very warm welcome and wished all success in achieving the objectives of the Workshop.

Dr George Pararas-Carayannis, Director of the International Tsunami Information Center (ITIC), was nominated and subsequently elected as Chairman. Mr Tom Sokolowski (U.S.A.) was designated as Rapporteur of the Workshop.

Dr Pararas-Carayannis welcomed participants and briefly reviewed the beneficial exchange of views and concepts between participants at the first International Workshop which took place in Sidney, BC, Canada, on 29–31 July 1985. The Chairman explained that the main idea of the Workshop is to bring together as many tsunami specialists from different countries as possible, and expressed the need to continue this exchange of views leading to better understanding of the practical needs for the mitigation of the tsunami hazard. He emphasized that this can be accomplished only with improved tsunami warning systems, better understanding of the latest scientific results for tsunami evaluation and prediction, and through a program of tsunami preparedness and public education. Then, he explained the rationale for the formulation of the Workshop program and the need by participants to review not only the tsunami threat in the Pacific Ocean, but in other world oceans and inland seas. Furthermore, he emphasized the need for Workshop participants to review their knowledge and information on existing seismic data processing systems, on data bases, latest developments on instrumentation, and on future projects, which may have practical application for tsunami disaster mitigation. The Chairman stated that tsunami disaster mitigation measures should be implemented, keeping in mind the objectives of the International Decade on Natural Disaster Reduction (IDNDR), a recent United Nations initiative.

A total of 28 presentations were made during the Workshop. The major subjects covered were the following:

1. International Cooperation in the Field of Tsunami Research and Warning
2. Survey of Existing Tsunami Warning Centers – Present Status, Results of Work, Plans for Future Development
3. Survey of Some Existing Seismic Data Processing Systems and Future Projects
4. Methods of Fast Evaluation of Tsunami Potential and Perspectives of their Implementation
5. Tsunami Data Bases
6. Tsunami Instrumentation and Observation
7. Tsunami Preparedness

Annex I provides the complete programme of the Workshop and the individual subjects covered. Following these presentations, recommendations were drafted and a general discussion ensued. The first recommendation addressed the need for cooperation between IUGG/Tsunami Commission and IOC/ITSU. The second recommendation dealt with the need for cooperation between ITSU and the Federation of Digital Broadband Seismograph Networks (FDSN) and outlined the expected benefits. The final recommendation addressed the need for tsunami warning systems in other areas of the globe in addition to the Pacific. After thorough discussion and deliberation, the final recommendations were prepared and adopted by the Group (Annex II). These recommendations were presented subsequently to the IUGG Tsunami Commission and to the ICG/ITSU Session and were discussed further in their agenda.

In conclusion, the second International IOC Tsunami Workshop was an overwhelming success. There were many valuable contributions that should result in material benefits to the International Tsunami Warning System in the Pacific and to the formation of regional tsunami warning systems in other vulnerable areas of the globe. Also, the Workshop brought into better focus the actions needed for implementation of the International Decade of Natural Disaster Reduction in dealing with the tsunami disaster. The Intergovernmental Oceanographic Commission (IOC), through its sponsorship, and the USSR Academy of Sciences' Computing Centers, through hosting and coordinating local arrangements, contributed greatly to the success of the Workshop.

GEORGE PARARAS-CARAYANNIS
Workshop Chairman
International Tsunami Information Center (U.S.A.)

ANNEX I: Programme of the Workshop

1. *Opening of the Workshop*

2. *International Cooperation in the Field of Tsunami Research and Warning*
   2.1. International Cooperation in the Field of Tsunami Research and Warning, (G. Pararas-Carayannis)

3. *Survey of Existing Tsunami Warning Centers – Present Status, Results of Work, Plans for Future Development*

   3.1. Pacific Tsunami Warning Center (G. Burton)
   3.2. Hawaii Regional Tsunami Warning System (G. Burton)
   3.3. Alaska Regional Tsunami Warning Center (T. Sokolowski)
   3.4. Japan Tsunami Warning Center (N. Hamada)
   3.5. USSR Tsunami Warning Center (B. Kuznetsov)
   3.6. French Polynesia Tsunami Warning Center (J. Talandier)
   3.7. Chile Tsunami Warning Center (E. Lorca)
   3.8. Tsunami Watches and Warnings in Fiji (G. Prasad)
   3.9. Assessment and Mitigation of the Tsunami Hazard in the Mediterranean Area (S. Tinti)

4. *Survey of Some Existing Seismic Data Processing Systems and Future Projects*
   4.1. Operative Seismic Data Processing in the NEIC and Plans for the New US National Seismic Network (J. Dewey)
   4.2. POSEIDON Project – its Application to the Better Understanding of Nature of the Interplate Earthquakes (R. Geller)

5. *Methods of Fast Evaluation of Tsunami Potential and Perspectives of their Implementation*
   5.1. A Review of Earthquake Prediction Methods (G. Pararas-Carayannis)
   5.2. Mm: A Variable-Period Mantle Magnitude (J. Talandier, E. Okal)
   5.3. On Earthquake Tsunami Generation Criteria (A. Ivashchenko, A. Poplavsky, S. Soloviev)

5.4. The Feasibility of Measuring the Low Frequency T Phase for Tsunami Warnings (S. Iwasaki)
5.5. Application of New Numerical Methods for Near-Real Time Tsunami Height Prediction (V. Gusiakov, A. Marchuk, V. Titov)
5.6. The Goal and Efficiency of the Automated Tsunami Warning System Project in the Far East of the U.S.S.R. (I. Kuzminykh, M. Malyshev, A. Metalnikov)
5.7. Integrated Warning System for Tsunami and Storm Surges in China (H. Yang)

6. *Tsunami Data Bases*
6.1. An Automated Tsunami Catalog. (A. Bobkov, C. Go, N. Zhigulina, K. Simonov)
6.2. Tsunami Data Base for British Columbia Tsunami Warnings. (T. Murty, W. Rapatz)
6.3. Historical Approach to the Study of Tsunamis: Recent U.S. Results (J. Lander)
6.4. The Development of Numerical Simulation of Tsunami Waves at the Computing Center at Krasnoyarsk (Yu. Shokin, L. Chubarov, V. Novikov, A. Sudakov, K. Simonov)

7. *Tsunami Instrumentation and Observation*
7.1. A Long-term Deep Tsunami Measurement Program: Strategy and Instrumentation (F. Gonzalez, E. Bernard, H. Milburn, D. Mattens)
7.2. Tsunami Observations Using Ocean Bottom Pressure Gauge (M. Okada, M. Katsumata)
7.3. Offshore Tsunami Warning Station – MEGA (G. Rybin)
7.4. Re-Use Plan of Commercial Submarine Communication Cable for Geophysical Research (J. Kasahara)

8. *Tsunami Preparedness*
8.1. Tsunamis of the 21st Century (G. Pararas-Carayannis)

9. *General Discussion and Adoption of Recommendations*

10. *Closure of the Session*

ANNEX II: Recommendations

Recommendation 1: Co-operation between IUGG/Tsunami Commission and IOC/ITSU

The Workshop,

Recognized that the majority of tsunami damages to human community occurs within 30 min and 400 km of its source;

Recognized further that sufficient scientific knowledge and technical expertise is currently available to develop appropriate early tsunami warning systems;

However, *recognized* also that many difficulties exist both in transferring scientific results to operational procedures and in communicating operational requirements to research communities;

Considered the objectives of the UN International Decade on Natural Disaster Reduction (IDNDR) and the need for international and interdisciplinary co-operation in mitigation of tsunami hazards;

Recommends that an *ad hoc* Joint IUGG/Tsunami Commission – IOC/ITSU Group of Experts be formulated with objectives of:

(i) formulating a project on tsunami disaster mitigation as a contribution to the International Decade on Natural Disaster Reduction;
(ii) providing adequate advice on implementation of the project to both sponsoring organizations;

Recommends also that IUGG and IOC seek possibility to hold a Joint Scientific and Technical Seminar on Mitigation of Tsunami Hazard in 1990/1991.

Recommendation 2: The need for co-operation between ITSU and the Federation of Digital Broadband Seismograph Networks (FDSN)

1. *Background*

(a) The tsunami community now recognizes the importance of broadband, wide-dynamic range seismic waveform data for issuing tsunami warnings. The tsunami community is therefore moving rapidly to establish real-time seismic networks.

(b) The international earthquake seismology community has established a consensus on the importance of broadband, wide-dynamic range seismic waveform data for studying: (i) the three dimensional distribution of elastic and anelastic properties of the earth's interior; and (ii) the details of the earthquake source process. Through IASPEI, the earthquake seismology community has established the FDSN for the purpose of (1) establishing standards for broadband seismic stations; (2) establishing formats and procedures for data exchange; and (3) co-ordinating the plans of various networks to avoid unnecessary duplication of effort.

(c) In general the members of ITSU have excellent liaison with earthquake seismologists in their own country. However, on an international level, at present, there is no formal liaison between ITSU and FDSN.

2. *Recommendations*

(a) Formal liaison between ITSU and FDSN should be established.

(b) When seismic stations are being established by ITSU Member States in support of Tsunami Warning Systems, FDSN Members should, whenever possible, be advised of this fact so that they may have the opportunity to investigate the feasibility of upgrading the stations to meet FDSN standards for broadband stations.*

3. *Expected Benefits*

(a) Both FDSN and ITSU will obviously benefit if the above recommendations are implemented.

(b) A real-time seismic network for tsunami warning also is a real-time network that permits accurate and almost instantaneous determination of the source parameters of *all damaging earthquakes*, anywhere in the world. This obviously is of tremendous importance for disaster relief authorities, and can greatly contribute to the goals of IDNDR.

Recommendation 3: Tsunami warning systems in other regions

Considering that tsunamis have occurred in the past in areas of the globe other than the Pacific also,

Considering that some of these tsunamis were reported to be highly disastrous resulting in great property damage and considerable catastrophic life loss,

Considering that growing world population, increasing urban concentration and larger investment in the infrastructure of societies are taking place nowadays particularly along the coastal regions and are expected to grow in the future,

Considering the important role played by ITSU towards international co-operation in tsunami research and tsunami warning systems,

Considering the important experience and achievement gained by actual ITSU member states and the needs to transfer such experience to other countries concerned with tsunami hazards.

It is recommended that ITSU urges the IOC to encourage the establishment of organizations similar to ITSU to address the needs of other tsunami-prone areas and to offer ITSU technical advice to these new organizations to facilitate their establishment and the development of Tsunami Warning Centers within their area of responsibility.

*Even if limitations on telemetry do not permit the transmission of the full bandwidth, the broadband data should be recorded on tape and sent to the data center rather than being discarded.

9 780792 311744